Fish Biotechnology

Fish Biotechnology

Dr. Sunita Rai

RANDOM

RANDOM PUBLICATIONS

NEW DELHI (INDIA)

Fish Biotechnology

ISBN 978-93-5111-517-5

Published in 2015 in India by

RANDOM PUBLICATIONS

4376-A/4B, Gali Murari Lal, Ansari Road
New Delhi-110 002
Phone : +9111-43580356, 011-23289044, 011-43142548
e-mail: sales@randompublications.com,
info@randompublications.com, randomexports@gmail.com

Reprinted 2026

Type Setting by : Friends Media, Delhi-110089

Preface

Demand for fish is soaring worldwide. It appears unlikely that the increasing demand can be met through increased natural harvest. There is international recognition that many of natural ocean and freshwater fisheries are being harvested to their limit. Aquaculture could help to meet increasing demand, and biotechnology can make a great contribution to improve aquaculture yields. Aquaculture animals are particularly well suited for research in biotechnology. Experimentation is facilitated by the availability of large numbers of gametes, use of external fertilization, and ease of in vitro rearing of embryos. In addition, many aquatic animals can be treated with hormones during development to induce sterility or functional sex reversal thus simplifying experimental procedures.

The agenda for modern biotechnology in aquaculture seems very similar to that of livestock and agriculture. Remarkable achievements have been made in the recent past in increasing production of crops, livestock and poultry through genetic and bio-technological tools. The potential areas of biotechnology in aquaculture include the use of synthetic hormones in induced breeding, production of monosex, uniparental and polyploid population, molecular biology, transgenic fish, gene banking, improved feeds and health management and development of natural products from marine organisms.

This book looks at a broad spectrum of biotech research efforts and their applications to the fisheries and aquaculture industry. It looks at the application of genetic, cellular, and molecular technologies to enable fish farmers to produce a more abundant, resilient, and healthier supply of seafood. It cover topics ranging from broodstock improvement to fish health and gene transfer. This book will be a valuable tool to researchers, producers, and students interested in better understanding the field of fish biotechnology.

Editor

Contents

1

Agricultural Applications of Biotechnology

Biotechnology is any technique that uses living organisms or substances from these organisms to make or modify a product for a practical purpose. Biotechnology can be applied to all classes of organism - from viruses and bacteria to plants and animals - and it is becoming a major feature of modern medicine, agriculture and industry. Modern agricultural biotechnology includes a range of tools that scientists employ to understand and manipulate the genetic make-up of organisms for use in the production or processing of agricultural products.

Some applications of biotechnology, such as fermentation and brewing, have been used for millennia. Other applications are newer but also well established. For example, micro-organisms have been used for decades as living factories for the production of life-saving antibiotics including penicillin, from the fungus *Penicillium*, and streptomycin from the bacterium *Streptomyces*. Modern detergents rely on enzymes produced via biotechnology, hard cheese production largely relies on rennet produced by biotech yeast and human insulin for diabetics is now produced using biotechnology.

Biotechnology is being used to address problems in all areas of agricultural production and processing. This includes plant breeding to raise and stabilize yields; to improve resistance to pests, diseases and abiotic stresses such as drought and cold; and to enhance the nutritional content of foods. Biotechnology is being used to develop low-cost disease-free planting materials for crops such as cassava, banana and potato and is creating new tools for the diagnosis and treatment of plant and animal diseases and for the measurement and conservation of genetic resources. Biotechnology is being used to speed up breeding programmes for plants, livestock and fish and to extend the range of traits that can be addressed. Animal feeds and feeding practices are being changed by biotechnology to improve animal nutrition and to reduce environmental waste. Biotechnology is used in disease diagnostics and for the production of vaccines against animal diseases.

Clearly, biotechnology is more than genetic engineering. Indeed, some of the least controversial aspects of agricultural biotechnology are potentially the most powerful and the most beneficial for the poor. Genomics, for example, is revolutionizing our understanding of the ways genes, cells, organisms and ecosystems function and is opening new horizons for marker-assisted breeding and genetic resource management. At the same time, genetic engineering is a very powerful tool whose role should be carefully evaluated. It is important to understand how biotechnology - particularly genetic engineering - complements and extends other approaches if sensible decisions are to be made about its use.

Managing Genetic Resources

Farmers and pastoralists have manipulated the genetic make-up of plants and animals since agriculture began more than 10 000 years ago. Farmers managed the process of domestication over millennia, through many cycles of selection of the best adapted individuals. This exploitation of the natural variation in biological organisms has given us the crops, plantation trees, farm animals and farmed fish of today, which often differ radically from their early ancestors.

The aim of modern breeders is the same as that of early farmers - to produce superior crops or animals. Conventional breeding, relying on the application of classic genetic principles based on the phenotype or physical characteristics of the organism concerned, has been very successful in introducing desirable traits into crop cultivars or livestock breeds from domesticated or wild relatives or mutants. In a conventional cross, whereby each parent donates half the genetic make-up of the progeny, undesirable traits may be passed on along with the desirable ones, and these undesirable traits may then have to be eliminated through successive generations of breeding. With each generation, the progeny must be tested for its growth characteristics as well as its nutritional and processing traits. Many generations may be required before the desired combination of traits is found, and time lags may be very long, especially for perennial crops such as trees and some species of livestock. Such phenotype-based selection is thus a slow, demanding process and is expensive in terms of both time and money. Biotechnology can make the application of conventional breeding methods more efficient.

Genomics

The most significant breakthroughs in agricultural biotechnology are coming from research into the structure of genomes and the genetic mechanisms behind economically important traits. The rapidly progressing discipline of genomics is providing information on the identity, location, impact and function of genes affecting such traits - knowledge that will increasingly drive the application of biotechnology in all agricultural sectors. Genomics sets the foundation for post-genomics activities, including new disciplines such as proteomics and metabolomics to generate knowledge on gene and protein structure, as well as their functions and interactions. These disciplines seek to understand systematically the molecular biology of organisms for their practical use.

A vast range of new and rapidly advancing technologies and equipment has also been developed to generate and process information about the structure and function of biological systems. The use and organization of this information is called bioinformatics. Advances in bioinformatics may allow the prediction of gene function from gene sequence data: from a listing of an organism's genes, it will become possible to build a theoretical framework of its biology. The comparison across organisms of physical and genetic maps and DNA sequences will significantly reduce the time needed to identify and select potentially useful genes.

Through the production of genetic maps that provide the precise location and sequences of genes, it is apparent that even distantly related genomes share common features. Comparative genomics assists in the understanding of many genomes based on the intensive study of just a few. For instance, the rice genome sequence is useful for studying the genomes of other cereals with which it shares features according to its degree of relatedness, and the mouse and malaria genomes provide models for livestock and some of the diseases that affect them. There are now model species for most types of crops, livestock and diseases and knowledge of their genomes is accumulating rapidly.

Molecular Markers

Reliable information on the distribution of genetic variation is a prerequisite for sound selection, breeding and conservation programmes. Genetic variation of a species or population can be assessed in the field or by studying molecular and other markers in the laboratory. A combination of the two approaches is required for reliable results. Molecular markers are identifiable DNA sequences, found at specific locations of the genome and associated with the inheritance of a trait or linked gene. Molecular markers can be used for (a) marker-assisted breeding, (b) understanding and conserving genetic resources and (c) genotype verification. These activities are critical for the genetic improvement of crops, forest trees, livestock and fish.

Marker-assisted Breeding

Genetic linkage maps can be used to locate and select for genes affecting traits of economic importance in plants or animals. The potential benefits of marker-assisted selection (MAS) are greatest for traits that are controlled by many genes, such as fruit yield, wood quality, disease resistance, milk and meat production, or body fat, and that are difficult, time-consuming or expensive to measure. Markers can also be used to increase the speed or efficiency of introducing new genes from one population to another, for example when wishing to introduce genes from wild relatives into modern plant varieties. When the desired trait is found within the same species, it may be transferred with traditional breeding methods, with molecular markers being used to track the desired gene.

Measuring and conserving genetic diversity

The use of molecular markers to measure the extent of variation at the genetic level, within and among populations, is of value in guiding genetic conservation activities and in the development of breeding populations in crops, livestock, forestry and fisheries. Studies carried out using these technologies in fish and forest tree species have revealed high levels of genetic variation both among and within populations. Livestock species are characterized by a high degree of genetic variation within populations, whereas crops exhibit a higher degree of variation across species. Data from other approaches, for example field observation, often cannot provide such information or are extremely difficult to collect.

Molecular markers are increasingly used to study the distribution and patterns of genetic diversity. Global surveys indicate, for example, that 40 percent of the remaining domestic livestock breeds are at risk of extinction. Most of these breeds are found only in developing countries, and there is often little knowledge about them or of their potential for improvement. They may contain valuable genes that confer adaptation or resilience to stresses, such as heat tolerance or disease resistance, that may be of use for future generations. Modern biotechnologies can help to counteract trends of genetic erosion in all food and agriculture sectors.

Genotype verification

Molecular markers have been widely used for identifying genotypes and for "genetic fingerprinting" of organisms. Genetic fingerprinting has been used in advanced tree-breeding programmes in which the correct identification of clones for large-scale propagation programmes is essential. Molecular markers have been used to identify endangered marine species that are either inadvertently captured in wild fisheries or that are purposefully taken illegally. Genotype verification is used intensively in parentage testing of domestic animals and for tracing livestock products in the food chain back to the farm and animal of origin.

Breeding and Reproducing of Crops and Trees

In addition to MAS, described above, a number of biotechnologies are used in breeding and reproducing crops and trees. Often these technologies are used in combination with each other and with conventional breeding approaches.

Cell and Tissue Culture and Micropropagation

Micropropagation involves taking small sections of plant tissue, or entire structures such as buds, and culturing them under artificial conditions to regenerate complete plants. Micropropagation is particularly useful for maintaining valuable plants, breeding otherwise difficult-to-breed species (e.g. many trees), speeding up plant breeding and providing abundant plant material for research. For crop and horticultural species, micropropagation is now the basis of a large commercial industry involving hundreds of laboratories around the world. In addition to its rapid propagation advantages, micropropagation can also be used to generate disease-free planting material, especially if combined with the use of disease-detection diagnostic kits. There have been some attempts to use micropropagation more widely in forestry. Compared with vegetative propagation through cuttings, the higher multiplication rates available through micropropagation offer a more rapid dissemination of planting stock, although limited availability of desirable clones is an impediment to its wider adoption in forestry.

In Vitro Selection

In vitro selection refers to the selection of germplasm by applying specific selection pressure to tissue culture under laboratory conditions. Many recent publications have reported useful correlations between *in vitro* responses and the expression of desirable field traits for crop plants, most commonly disease resistance. Positive results are available also for tolerance to herbicides, metals, salt and low temperatures. For the selection criteria of major general importance in forest trees (in particular vigour, stem form and wood quality), poor correlations with field responses still limit the usefulness of *in vitro* selection. However, this method may be of interest in forestry programmes for screening disease resistance and tolerance to salt, frost and drought.

Genetic Engineering

When the desired trait is found in an organism that is not sexually compatible with the host, it may be transferred using genetic engineering. In plants, the most common method for genetic engineering uses the soil bacterium *Agrobacterium tumefasciens* as a vector. Researchers insert the desired gene or genes into the bacterium and then infect the host plant. The desired genes are transmitted to the host along with the infection. This method is used mainly with dicot species such as tomato and potato. Some crops, particularly monocot species such as wheat and rye, are not naturally susceptible to transformation via *A. tumefasciens*, although the method has recently been successfully used to transform wheat and other cereals. In the most common transformation technique for these crops, the desired gene is coated on gold or tungsten particles and a "gene gun" is used literally to shoot the gene into the host at high velocity.

Three distinctive types of genetically modified crops exist: (a) "distant transfer", in which genes are transferred between organisms of different kingdoms (e.g. bacteria into plants); (b) "close transfer", in which genes are transferred from one species to another of the same kingdom (e.g. from one plant to another); and (c) "tweaking", in which genes already present in the organism's genome are manipulated to change the level or pattern of expression. Once the gene has been transferred, the crop must be tested to ensure that the gene is expressed properly and is stable over several generations of breeding. This screening can usually be performed more efficiently than for conventional crosses because the nature of the gene is known, molecular methods are available to determine its localization in the genome and fewer genetic changes are involved.

Most of the transgenic crops planted so far have incorporated only a very limited number of genes aimed at conferring insect resistance and/or herbicide tolerance. However, some transgenic crops and traits of greater potential interest for developing countries have been developed but have not yet been released commercially.

Nutritionally enhanced crops could make a significant contribution to the reduction of micronutrient malnutrition in developing countries. Biofortification (the development of nutritionally enhanced foods) can be advanced through the application of several biotechnologies in combination. Genomic analysis and genetic linkage mapping are needed to identify the genes responsible for natural variation in nutrient levels of common foods. These genes can then be transferred into familiar cultivars through conventional breeding and MAS or, if sufficient natural variation does not occur within a single species, through genetic engineering. Non-transgenic approaches are being used, for example, to enhance the protein content in maize, iron in rice, and carotene in sweet potato and cassava.

Genetic engineering can be used when insufficient natural variation in the desired nutrient exists within a species. The well-known transgenic Golden Rice contains three foreign genes - two from the daffodil and one from a bacterium - that produce provitamin A. Scientists are well on their way to developing transgenic "nutritionally optimized"' rice that would contain genes producing provitamin A, iron and more protein. Other nutritionally enhanced foods are under development, such as oils with reduced levels of undesirable fatty acids. In addition, foods that are commonly allergenic (shrimp, peanuts, soybean, rice, etc.) are being modified to contain lower levels of allergenic compounds.

A major technical factor limiting the application of genetic modification to forest trees is the current low level of knowledge regarding the molecular control of traits that are of most interest. One of the first reported trials with genetically modified forest trees was initiated in Belgium in 1988 using poplars. Since then, there have been more than 100 reported trials involving at least 24 tree species, primarily timber-producing species. Traits for which genetic modification has been contemplated for forest trees include insect and virus resistance, herbicide tolerance and lignin content. Reduction of lignin is a valuable objective for species producing pulp for the paper industry because it would enable a reduction in the use of chemicals in the process.

Breeding and Reproducing Livestock and Fish

Biotechnology has long been a source of innovation in livestock and aquaculture production and processing and has had a profound impact on both sectors. Rapid advances in molecular biology and further developments in reproductive biology provide new and powerful tools for further innovation.

Technologies such as genomics and molecular markers, as described above, are valuable in understanding, characterizing and managing genetic resources in livestock and fisheries as well as in crops and forestry. Genetic engineering is also relevant in livestock and fisheries, although the techniques differ, and additional reproductive technologies are available in these sectors. This section describes the reproductive biotechnologies that are specific to the livestock and fisheries sectors.

The main objective of reproductive biotechnologies for livestock is to increase reproductive efficiency and rates of animal genetic improvement. The genetic improvement of locally adapted breeds will be important in realizing sustainable production systems within the broad spectrum of developing country production environments, and will probably best be realized by the strategic use of both non-genetic and genetic interventions. Reproductive biotechnology in fisheries presents opportunities to increase growth rates and improve the management of farmed species and to limit the reproductive potential of genetically engineered species.

Artificial Insemination and Multiple Ovulation/Embryo Transfer

Advances in artificial insemination (AI) and multiple ovulation followed by embryo transfer (MOET) have already had a major impact on livestock improvement programmes in developed countries and many developing countries because they speed up the process of genetic improvement, reduce the risk of disease transmission and expand the number of animals that can be bred from a superior parent - the male in the case of AI and the female in the case of MOET. They also increase the incentives for private research in animal breeding and significantly expand the market for improved parent stock.

The number of AIs performed globally during 1998 was over 100 million in cattle (primarily dairy cattle, including buffalo), 40 million in pigs, 3.3 million in sheep and 0.5 million in goats. These figures illustrate both the higher economic returns in dairy cattle and the fact that cattle semen is much easier to deep-freeze than semen from other animals. Although over 60 million cattle AIs were performed in South and Southeast Asia, fewer than 1 million were performed in Africa.

AI is only effective when the farm sector has access to considerably greater technical and institutional and other organizational capacity than when male animals are used directly for breeding purposes. On the positive side, farmers employing AI do not have to face the costs or hazards of rearing breeding males and can have access to semen from any part of the world.

Despite the widespread use of AI in developed countries and in many developing countries, including within more advanced smallholder systems, it is applied only on farms that practise intensive livestock management with high-value animals. This is clearly not because of technical problems with semen production and storage, as most procedures are now fully standardized and proven to be effective under tropical developing-country conditions. Rather, it is because of the many organizational, logistical and farmer-training constraints that influence the quality and efficiency of the technology.

MOET takes AI one step further, both in terms of genetic gains possible and level of technical capacity and organization required. MOET is one of the basic technologies for the application of more advanced reproductive biotechnologies such as cloning and transgenics. During 2001 the number of embryos transferred globally was 450 000, mainly in dairy cattle, with

North America and Europe accounting for 62 percent, followed by South America (16 percent) and Asia (11 percent). About 80 percent of the bulls used in AI are derived from MOET. The main potential advantage of MOET for developing countries will be in the possibility of importing frozen embryos instead of live animals, for example in the establishment of nucleus breeding stocks of locally adapted genetic resources, with the related lower sanitary risks.

Chromosome-set Manipulation and Sex Reversal in Fish

Controlling the sex and reproductive capacity of fish can be important for commercial and environmental reasons. One sex is often more desirable than the other; for example, only female sturgeon produce caviar and male tilapia grow faster than females. Sterility may be desirable when reproduction affects the taste of the product (e.g. oysters) or when farmed species (transgenic or not) might breed with wild populations. Chromosome-set manipulation and sex reversal are well-established techniques to control these factors. In chromosome-set manipulation, temperature, chemical and pressure shocks applied to fish eggs can be used to produce individuals that have three sets of chromosomes rather than the usual two. These triploid organisms generally do not channel energy into reproduction and thus are functionally sterile. Sex reversal can be accomplished by several methods including administering appropriate hormones. For example, genetically male tilapia can be turned into females through oestrogen treatments. These genetic males, when mated with normal males, produce a group of all-male tilapia.

Genetic Engineering in Livestock and Fish

Genetic engineering in animals can be used to introduce foreign genes into the animal genome or, alternatively, to "knock-out" selected genes. The method most used at present involves direct microinjection of DNA into the pronuclei of fertilized eggs, but progress is being made with new approaches such as nuclear transfer and the use of lentiviruses as DNA vectors.

In the first genetic engineering experiments with farm animals, genes responsible for growth were introduced into pigs to increase growth and improve carcass quality. Current research efforts include engineering resistance to animal diseases, such as Marek's disease in poultry, scrapie in sheep and mastitis in cattle, and diseases that affect human health such as*Salmonella* in poultry. Other examples include increasing the casein content of milk and inducing the production of pharmaceutical or industrial chemicals in the milk or semen of animals. Although conceptually simple, the methods used to genetically engineer livestock require special equipment and considerable dexterity, and no agricultural applications have proved commercially successful thus far. Applications in the near future therefore seem to be limited to the production of transgenic animals for use in the production of industrial or pharmaceutical products.

Genetic engineering is an active area of research and development in aquaculture. The large size and hardy nature of many fish eggs allow them to be manipulated easily and facilitate gene transfer by direct injection of a foreign gene or by electroporation, in which an electric field assists gene transfer. Gene transfer in fish has usually involved genes that produce growth hormone and has been shown to increase growth rates dramatically in carp, salmon, tilapia and other species. In addition, a gene from the winter flounder that produces an antifreeze protein was put into salmon in the hope of extending the farming range of the fish. The gene did not produce enough of the protein to extend the salmon's range into colder waters, but it did allow the salmon to continue growing during cold months when non-transgenic salmon would not grow. These applications are still in the research and development stage, and no transgenic aquatic animals are currently available to the consumer.

OTHER BIOTECHNOLOGIES

Diagnostics and Epidemiology

Plant and animal diseases are difficult to diagnose because the signs may be misleading or even entirely absent until serious damage has occurred. Advanced biotechnology-based diagnostic tests make it possible to identify disease-causing agents and to monitor the impact of disease control programmes to a degree of precision not previously possible. Molecular epidemiology characterizes pathogens (viruses, bacteria, parasites and fungi) by nucleotide sequencing, which enables their origin to be traced. This is particularly important for epidemic diseases, in which the possibility of pinpointing the source of infection can significantly contribute to improved disease control. For example, the molecular analysis of rinderpest viruses has been vital for determining the lineages circulating in the world and instrumental in aiding the Global Rinderpest Eradication Programme (GREP). Enzyme-linked immunosorbent assay (ELISA) tests have become the standard methodology for the diagnosis and surveillance of many animal and fish diseases worldwide, and the polymerase chain reaction (PCR) technique is especially useful in diagnosing plant diseases and is proving increasingly so also for livestock and fish diseases. The effectiveness of plant and animal health programmes is also being considerably enhanced by the development of genetic probes that allow specific pathogens to be distinguished and detected in tissue, whole animals and even in water and soil samples.

Vaccine Development

Genetically engineered vaccines are being developed to protect fish and livestock against pathogens and parasites. Although vaccines developed using traditional approaches have had a major impact on the control of foot-and-mouth and tick-borne diseases, rinderpest and other diseases affecting livestock, recombinant vaccines can offer various advantages over conventional vaccines in terms of safety, specificity and stability. Importantly, such vaccines, coupled with the appropriate diagnostic test, allow the distinction between vaccinated and naturally infected animals. This is important in disease control programmes as it enables continued vaccination even when the shift from the control to the eradication stage is contemplated.

Today, quality improved vaccines are available for, for example, Newcastle disease, classical swine fever and rinderpest. In addition to the technical improvements, advances in biotechnology will make vaccine production cheaper, and therefore improve supply and availability for smallholders.

Animal Nutrition

Biotechnologies have already resulted in animal nutrition aids such as enzymes, probiotics, single-cell proteins and antibiotic feed additives that are already widely used in intensive production systems worldwide to improve the availability of nutrients from feeds and the productivity of livestock and aquaculture. Gene-based technologies are being increasingly employed to improve animal nutrition, either through modifying the feeds to make them more digestible or through modifying the digestive and metabolic systems of animals to enable them to make better use of the available feeds. Although progress in the latter approach is likely to be slow because of gaps in our current understanding of the underlying genetics, physiology and biochemistry, one example of commercial success in high-input, intensively managed systems is the use of recombinant somatotropin, a hormone that results in increased milk production in dairy cows and accelerated growth and leaner carcasses in meat animals.

Future Concerns

Biotechnology is a complement - not a substitute - for many areas of conventional agricultural research. It offers a range of tools to improve our understanding and management of genetic resources for food and agriculture. These tools are already making a contribution to breeding and conservation programmes and to facilitating the diagnosis, treatment and prevention of plant and animal diseases. The application of biotechnology provides the researcher with new knowledge and tools that make the job more efficient and effective. In this way, biotechnology-based research programmes can be seen as a more precise extension of conventional approaches. At the same time, genetic engineering can be seen as a dramatic departure from conventional breeding because it gives scientists the power to move genetic material between organisms that could not be bred through classical means.

Agricultural biotechnology is cross-sectoral and interdisciplinary. Most of the molecular techniques and their applications are common across all sectors of food and agriculture, but biotechnology cannot stand on its own. Genetic engineering in crops, for example, cannot proceed without knowledge derived from genomics and it is of little practical use in the absence of an effective plant-breeding programme. Any single research objective requires mastery of a bundle of technological elements. Biotechnology should be part of a comprehensive, integrated agricultural research programme that takes advantage of work in other sectoral, disciplinary and national programmes. This has broad implications for developing countries and their development partners as they design and implement national research policies, institutions and capacity-building programmes.

Agricultural biotechnology is international. Although most of the basic research in molecular biology is taking place in developed countries, this research can be beneficial for developing countries because it provides insight into the physiology of all plants and animals. The findings of the human and the mice genome projects provide direct benefits for farm animals, and vice versa, whereas studies of maize and rice can provide parallels for applications in subsistence crops such as sorghum and tef. However, specific work is needed on the breeds and species of importance in developing countries. Developing countries are host to the greatest array of agricultural biodiversity in the world, but little work has been done on characterizing these plant and animal species at the molecular level to assess their production potential and their ability to resist disease and environmental stresses or to ensure their long-term conservation.

The application of new molecular biotechnologies and new breeding strategies to the crops and livestock breeds of specific relevance to smallholder production systems in developing countries will probably be constrained in the near future for a number of reasons. These include lack of reliable longer-term research funding, inadequate technical and operational capacity, the low commercial value of the crops and breeds, lack of adequate conventional breeding programmes and the need to select in the relevant production environments. Nevertheless, developing countries are already faced with the need to evaluate genetically modified (GM) crops and they will one day also need to evaluate the possible use of GM trees, livestock and fish. These innovations may offer opportunities for increased production, productivity, product quality and adaptive fitness, but they will certainly create challenges for the research and regulatory capacity of developing countries.

References

Carpenter, J.E. & Gianessi, L.P. 2001. *Agricultural biotechnology: updated benefits estimates*. Washington, DC, National Center for Food and Agricultural Policy.

Hoban, T. 2004. *Public attitudes towards agricultural biotechnology*. ESA Working Paper, forthcoming. Rome, FAO.

Pingali, P & Raney, T. 2003. *Globalization and agricultural biotechnology: impacts and implications for developing countries*. ESA Working Paper. Rome, FAO.

Pray, C.E. & Naseem, A. 2003. *Biotechnology R&D: policy options to ensure access and benefits for the poor*. ESA Working Paper 03-08. Rome, FAO.

Thompson, P.B. 1997. *Food biotechnology in ethical perspective*. London, Blackie Academic & Professional.

Traxler, G. 2004. *Economic impacts of biotechnology-based technological innovations*. ESA Working Paper, forthcoming. Rome, FAO.

2

Techniques of Animal Biotechnology

Animal biotechnology is the application of scientific and engineering principles to the processing or production of materials by animals or aquatic species to provide goods and services. Examples of animal biotechnology include generation of transgenic animals or transgenic fish (animals or fish with one or more genes introduced by human intervention), using gene knockout technology to generate animals in which a specific gene has been inactivated, production of nearly identical animals by somatic cell nuclear transfer or production of infertile aquatic species.

Given the breadth of the term "animal biotechnology," one might reasonably include thousands of years of humans selectively breeding animals as a biotechnology: Observing desirable animal traits and attempting to breed those traits into successive lines of animals. Artificial insemination (AI) in the 1950s was a technological advance in traditional selective breeding and an important adjunct to the development of modern industrial animal production, especially in dairy and poultry. AI was adopted by producers and accepted by the public with virtually no controversy.

Estrus synchronisation, a related technology that improved the efficiency of AI by more accurately controlling when a female was in heat, is also an important animal biotechnology. With the development in the 1970s and patenting in the 1980s of recombinant DNA techniques, and the subsequent analysis of genes, their resulting proteins, and the role played by the proteins in animal biochemical processes (functional genomics), modern biotechnology is increasingly equipped with a set of sophisticated tools holding the promise of transforming the selective breeding of animals. The range of new techniques and technologies is only now beginning to transform animal biotechnology in ways that plant biotechnology was transformed in the 1980s and 1990s.

Modern animal biotechnology is developing against the background of public experience with plant biotechnology, and controversy over the technologies may be a continuing feature of animal biotechnology development, not least because of the closer connection between humans and some animals and the belief that techniques developed for animals are only a step away from application to humans. Some of the better known animal biotechnologies include:

— *Embryo transfer.* After AI and estrus synchronisation, embryo transfer (ET) is the third most commonly used biotechnology. In ET, a donor cow of superior breeding is chemically induced to superovulate. The eggs are then fertilised within the donor, the embryo develops and is then removed and implanted in a recipient cow. Between removal and implantation, embryos may be frozen for safekeeping. Because of the relatively high costs, ET is used mostly within registered cowherds.

— *In vitro fertilisation.* With in vitro fertilisation (IVF), a technician removes unfertilised eggs (oocytes) from the donor cow's ovaries, usually recovering 6-8 useable oocytes. The oocytes mature in an incubator and are fertilised with sperm. The resulting zygotes incubate and develop in the laboratory before being placed into the recipient cow. While IVF can produce many fertilised embryos, the added expense of ET makes the procedure prohibitive in most cases.

— *Sexing embryos.* The dairy industry prefers heifers and the beef industry prefers bulls. Embryo sexing methods in cattle have been developed using a bovine Y-chromosome probe. Technicians remove a few cells from the embryo and assess the DNA in these cells for the presence of a Y-chromosome. Presence of a Ychromosome determines the embryo is male. Research is also developing in sperm sexing technology.

— *Transgenics.* A prominent area of contemporary animal biotechnology research is the development of transgenic animals through genetic engineering (GE) technology.Transgenic animals are produced by introducing an isolated DNA fragment into an embryo so that the resulting animal will express a desired trait.The only current routine use of transgenic animals, primarily mice, is in the area of human disease research. Potential agricultural applications for such genetic engineering, however, include improved feed use and faster growth; more resistance to disease; meat that is leaner or that has more of some other desirable quality; and possibly even animal waste that is more environmentally benign.

— *Cloning.* Cloning is a biotechnology developing rapidly and with significant public controversy. Most people think of cloning as the creation of an organism that is genetically identical to another one. However, scientists use the term more broadly, to refer to production not only of such organisms but also genetically identical cells, and to replication of DNA and other molecules. It also refers to a form of reproduction found naturally in many single-celled organisms, as well as plants and animals. Those differences in meaning and usage have caused some confusion in public debate about cloning, where the main area of genuine controversy relates to artificial cloning involving higher organisms, including humans.

Embryo Transfer

Embryo transfer refers to a step in the life and times of in vitro fertilisation (IVF) whereby one or several embryos are placed into the uterus of the female with the intent to establish a pregnancy. Embryos can be either "fresh" from fertilised egg cells of the same menstrual cycle, or "frozen", that is they have been generated in a preceding cyle, cryopreserved, and are thawn just prior to the transfer.

In the human, the uterine lining (endometrium) needs to be appropriately prepared so that the embryo(s) can implant. In a natural or stimulated cycle, the embryo transfer takes place in the luteal phase at a time where the lining is appropriately developed in relation to the status of the embryo. In a cycle where a "frozen" embryo is transferred, the recipient woman could be given first estrogen preparations (about 2 weeks), then a combination of estrogen and progesterone so that the lining becomes receptive for the embryo. The time of receptivity is the "implantation window".

In stimulated cycles in human IVF, embryos are typically transferred 3 days after fertilisation and may then be at the eight-cell stage, or they are transferred 2 to 3 days later when they have reached the blastocyst stage. Embryos who reach the day 3 cell stage can be tested for chromosal or specific genetic defects prior to possible transfer by preimplantation genetic diagnosis (PGD).

The procedure of embryo transfer is performed by a physician, often with the aid of ultrasound to allow for precise placement. The catheter loaded with one or more embryos is advanced through the cervix into the uterine cavity and the embryo(s) are released into the cavity. Anesthesia is generally not required.

A major issue is how many embryos should be transferred. Placement of multiple embryos carries the risk of multiple pregnancy. In the past, physicans have often placed too many embryos in the hope to establish a pregnancy. However, the rise in multiple pregnancies has led to a reassessment of this approach. Professional societies and in many countries, the legislature, have issued guidelines or laws to curtail a practice of placing too many embryos in an attempt to reduce multiple pregnancies. After embryo transfer patients are kept on estrogen and progesterone medication; pregnancy testing is done typically 2 weeks later.

It is not necessary that the embryo transfer be performed on the female who provided the eggs. Thus another female whose uterus is appropriately prepared can receive the embryo and carry the pregnancy. This situation may apply for a woman who has eggs but no uterus and wants to have a biologic baby; she would require the help of a gestational carrier to carry the pregnancy. Also, a woman who has no eggs but a uterus may resort to egg donor IVF, in which case another woman would provide eggs for fertilisation and the resulting embryos are placed into the uterus of the patient. Third party reproduction is controversial and regulated in many countries.

In Vitro Fertilisation

In vitro fertilisation (IVF) is a technique in which egg cells are fertilised by sperm outside the woman's womb. IVF is a major treatment in infertility when other methods of achieving conception have failed. The process involves hormonally controlling the ovulatory process, removing ova (eggs) from the woman's ovaries and letting sperm fertilise them in a fluid medium. The fertilised egg (zygote) is then transferred to the patient's uterus with the intent to establish a successful pregnancy.

Initially IVF was developed to overcome infertility due to problems of the fallopian tube, but it turned out that it was successful in many other infertility situations as well. The introduction of intracytoplasmic sperm injection (ICSI) addresses the problem of male infertility to a large extent. Thus, for IVF to be successful it may be easier to say that it requires healthy ova, sperm that can fertilise, and a uterus that can maintain a pregnancy. Cost considerations generally place IVF as a treatment when other less expensive options have failed. This means that IVF can be used for females already gone through pregnancy. The donated oocyte can be fertilised in a crucible. If the fertilisation is successful, the fertilised egg will be transferred into the uterus, within which it will develop into an embryo.

Treatment cycles are typically started on the third day of menstruation and consist of a regimen of fertility medications to stimulate the development of multiple follicles of the ovaries. In most patients injectable gonadotropins (usually FSH analogues) are used under close monitoring. Such monitoring frequently checks the oestradiol level and, by means of gynaecologic ultrasonography, follicular growth. Typically approximately 10 days of injections will be necessary. Endogenous ovulation is blocked by the use of GnRH agonists or GnRH antagonists.

When follicular maturation is judged to be adequate, human chorionic gonadotropin is given. This agent, which acts as an analogue of luteinising hormone, would cause ovulation about 36 hours after injection, but a retrieval procedure takes place just prior to that, in order to recover the egg cells from the ovary. The eggs are retrieved from the patient using a transvaginal technique involving an ultrasound-guided needle piercing the vaginal wall to reach the ovaries. Through this needle follicles can be aspirated, and the follicular fluid is handed to the IVF laboratory to identify ova. The retrieval procedure takes about 20 minutes and is usually done under conscious sedation or general anaesthesia.

In the laboratory, the identified eggs are stripped of surrounding cells and prepared for fertilisation. In the meantime, semen is prepared for fertilisation by removing inactive cells and seminal fluid. The sperm and the egg are incubated together (at a ratio of about 75,000:1) in the culture media for about 18 hours. By that time fertilisation should have taken place and the fertilised egg would show two pronuclei. In situations where the sperm count is low a single sperm is injected directly into the egg using intracytoplasmic sperm injection (ICSI). The fertilised egg is passed to a special growth medium and left for about 48 hours until the egg has reached the 6-8 cell stage.

Laboratories have developed grading methods to judge oocyte and embryo quality. Typically, embryos that have reached the 6-8 cell stage are transferred three days after retrieval. In many American programmes, however, embryos are placed into an extended culture system with a transfer done at the blastocyst stage, especially if many good-quality day-3 embryos are available. Blastocysts may give higher pregnancy rates if embryo quality is in doubt. However, many studies have shown no difference in pregnancy rates between day-3 and day-5 transfers. In Europe, day-2 transfers are common.

Embryos are graded by the embryologist based on the number of cells, evenness of growth and degree of fragmentation. The number to be transferred depends on the number available, the age of the woman and other health and diagnostic factors. In countries such as the UK and New Zealand, a maximum of two embryos are transferred except in unusual circumstances. This is to limit the number of multiple pregnancies. The embryos judged to be the "best" are transferred to the patient's uterus through a thin, plastic catheter, which goes through her vagina and cervix. Often, several embryos are passed into the uterus to improve chances of implantation and pregnancy.

The patient has to wait two weeks before she returns to the clinic for the pregnancy test. During this time she may receive progesterone—a hormone that keeps the uterus lining thickened and suitable for implantation. Many IVF programmes provide additional medications as part of their protocol.

Chance of a successful pregnancy is approximately 20-30% for each IVF cycle, although selected clinics are now able to quote rates up to 50% per cycle. There are many factors that determine success rates including the age of the patient, the quality of the eggs and sperm, the duration of the infertility, the health of the uterus, and the medical expertise. It is a common practice for IVF programmes to boost the pregnancy rate by placing multiple embryos during embryo transfer. A flip side of this practice is a higher risk of multiple pregnancy, itself associated with obstetric complications.

IVF programmes generally publish their pregnancy rates. However, comparisons between clinics are difficult as many variables determine outcome. Furthermore, these statistics depend strongly on the type of patients selected.

There are many reasons why pregnancy may not occur following IVF and embryo transfer, including:

— The timing of ovulation may be misjudged, or ovulation may not be able to be predicted or may not occur
— Attempts to obtain eggs that develop during the monitored cycle may be unsuccessful
— The eggs obtained may be abnormal or may have been damaged during the retrieval process
— A semen specimen may not be able to be provided
— Fertilisation of eggs to form embryos may not occur
— Cleavage or cell division of the fertilised eggs may not take place
— The embryo may not develop normally
— Implantation may not occur
— Equipment failure, infection and/or human error or other unforeseen and uncontrollable factors, which may result in the loss of or damage to the eggs, the semen sample and/or the embryos

Sexing of Embryos

Sexing of embryos before transfer and implanting has great potential for the livestock industry. Manipulating the sex of offspring has been a dream of the cattle industry for decades. A number of approaches to the sexing of semen have been attempted, and several have been reported as successful. However, the only method of semen sexing that has shown any promise has been the sorting of spermatozoa according to the DNA content, by means of flow cytometry. This methodology is severely limited by the small number of cells that can be sorted per hour and the high cost of the equipment.

The widespread use of embryo transfer technology has created an alternative approach to controlling the sex of offspring. Embryo sexing has been attempted by a variety of methods, including cytogenetic analysis, assays for X-linked enzyme activity, analysis of differential development rates, detection of male-specific antigens, and the use of Y-specific DNA probes. In DNA amplification techniques, the probe itself is not used. The sequence information generated from the identification and cloning of the probe is used instead.

Since the development of DNA amplification techniques, particularly the polymerase chain reaction (PCR), these techniques have been applied to numerous situations in which the analysis of rare sequences is desired. It was the identification of bovine Y-chromosome specific DNA probes, and the subsequent development of DNA amplification techniques by PCR, that made the possibility of sexing embryos into a reality.

The best donor in embryo transfer (ET) is generally a young cow, lactating but not at the peak lactation stage, with no history of reproductive problems. Too much body fat will decrease the success of ET. Fat cows generally respond poorly to superovulation treatments. A suitable recipient is a heifer or a cow that will cycle normally and have a good conception rate. At the time of ET, the recipient must be in adequate heat synchrony with the donor, and have a palpable corpus luteum.

The most common method used is a two-shot prostaglandin regime. Prostaglandin is given twice, with the shots eleven days apart. The second injection is timed two days before the donor is due to come into heat. Recipients take 2-3 days to come into heat after prostaglandin, if they respond at all.

Superovulation is the process of stimulating multiple follicular development and ovulation by administering gonadotrophic hormones, followed by a prostaglandin injection. There are two main gonadotrophins in common use for the superovulation of donor cows. PMSG has a half-life of up to five days, and is administered as a single injection. FSH-P has a half-life of 2 - 5 hours, and is administered twice daily for 3.5 days. Heat should occur 48 hours after the PG injection. PG may be given on the third day of a 3.5-day FSH-P stimulation. A total dosage of 36-40 mg of FSH-P with a decline dose is administered.

At the present time, non-surgical embryo recovery is the technique in world-wide use. The base of the tail of the donor is clipped, and an epidural injection of 6 mL of 2% procaine is given. The French IMV three-way collection equipment with an extendible head catheter then is manipulated into the selected horn, so that the inflatable balloon is in the right place. The balloon is inflated quickly with 20 mL of air. Approximately 600 to 800 mL of Dulbecco's phosphate buffered saline, plus 1% bovine serum medium warmed to 37° C, are placed for infusion.

It is important that the culture medium and all equipment used to handle the embryos is sterile. The embryos are frozen in a medium containing

Dulbecco's phosphate buffered saline solution with 1.5M ethylene glycol, 0.4% BSA and 0.1M sucrose. The freezing procedure is as follows. Embryos are put into a 1.5M EG solution, and loaded into a 0.25 mL straw. After five minutes, they are placed in a 6°C freezing chamber, and kept at -6°C for 10 minutes. The cooling rate then continues at 0.5°C/minute until a temperature of -32°C is reached. The embryo is then plunged into liquid nitrogen.

The direct transfer procedure is as follows. The straw is removed from the liquid nitrogen, and held in the air for five seconds. It is then immersed in a water bath at 25°C for 25 seconds. The straw is then removed, wiped dry, loaded into an ET gun, and transferred directly into the cow. For optimal results, the transfer should take place within six minutes as the embryo is loaded into the ET gun.

Transgenics

Advances in scientific discovery and laboratory techniques in the last half of the twentieth century resulted in the ability to manipulate the deoxyribonucleic acid of organisms and gave rise to transgenic animals. Since the early 1980s, methods have been developed and refined to generate transgenic animals or transgenic aquatic species. For example, transgenic livestock and transgenic aquatic species have been generated with increased growth rates, enhanced lean muscle mass, enhanced resistance to disease or improved use of dietary phosphorous to lessen the environmental impacts of animal manure. Transgenic poultry, swine, goats, and cattle also have been produced that generate large quantities of human proteins in eggs, milk, blood, or urine, with the goal of using these products as human pharmaceuticals. Examples of human pharmaceutical proteins include enzymes, clotting factors, albumin, and antibodies. The major factor limiting widespread use of transgenic animals in agricultural production systems is the relatively inefficient rate of production of transgenic animals.

The use of transgenic animals may accelerate classical breeding programmes and provide a means for the economical production of life-saving pharmaceuticals. To understand how a transgenic animal is produced, it is necessary to the basic components and functions of living organisms. Animals are made of billions of cells all working together. Every cell of the animal has a complete "instruction manual" or genome that is inherited from the parents of the animal as a combination of their genomes. The genome resides in the nucleus of the cell. Genes are found within the genome and serve as the "words" of the instruction manual. When a cell reads a word, a specific protein is produced. Proteins give an individual cell, and therefore the animal, its form and function. Genes are written using the four-letter alphabet A, C, G, T. The letters stand for four chemicals called bases, which together compose DNA. DNA is universal in nature, meaning that the four chemical bases of DNA are the same in all living organisms. Consequently, a gene from one organism can function in another organism. When a new gene is put into an animal's genome it is said to be transgenic. Scientists have also discovered how to genetically modify when and/or where in an animal a gene is expressed. For example, it is possible to produce new proteins in the milk of a cow and nowhere else in its body.

One way to produce transgenic animals is through a technique called microinjection. Once scientists have identified and isolated the piece of DNA comprising the gene to be transferred, it is injected into a fertilised egg of the desired animal using a very small glass needle visualised under a microscope. In approximately one percent of the injected eggs, the gene becomes a new "word" in the egg's "instruction manual" by physically combining with the egg's genome. Ideally, the new gene integrates into the genome before the egg begins to divide. If this occurs, every cell in the animal can contain the new protein and the animal will pass the gene on to its offspring. After injection of the gene, the fertilised egg is implanted into a surrogate mother where it fully develops into a transgenic animal.

Currently, the only routine commercial use of transgenic animals (primarily mice) is in the area of human disease research. One way to characterise the range of genetic modifications that are being considered for use in animals is in the three broad areas of input, output, and value-added traits. Examples of each are described below.

— *Input traits*. An "input" trait helps livestock and dairy producers by increasing production efficiency. Input traits that are being investigated for use in animals:
    — Faster, more efficient growth rates
    — Increased production of milk or wool
    — Resistance to diseases caused by viruses and bacteria
— *Output traits*. An output trait helps consumers or downstream processors by enhancing the quality of the animal product. Output traits that may prove to be beneficial:

- — Leaner, more tender beef and pork
- — Milk that lacks allergenic proteins, or results in increased amounts of cheese and yogurt

— *Value-added traits*. By adding or modifying genes, animals can function in completely new ways.

- — Producing large amounts of therapeutic proteins in animal milk may be an efficient, relatively low cost method to manufacture many proteins used to treat human diseases or proteins that have industrial value.
- — Transplanting animal organs into humans, or xenotransplantation, can be made more successful by genetically modifying the organs so that they are not as readily rejected by the human immune system.
- — Development of animals that serve as models for human diseases to help scientists better understand prevention and treatment strategies.

One concern of transgenic animal technology is the welfare of the animals. Developmental and health abnormalitites have been reported in conjunction with its use; therefore, researchers must take care to minimise animal suffering. The inadvertant release or escape of transgenic animals into the wild where they could breed or compete with the natural population is often cited as a potential risk to the environment. The actual risk associated with this will depend on the type of animal and the nature of the genetic modification; however, where appropriate, procedures must be in place to alleviate this concern.

Aquaculture produces approximately one-third of all fish and shellfish consumed by humans. Researchers are working with transgenic catfish, carp, tilapia, striped bass, clams, oysters, shrimp, and abalone. Traits being tested in transgenic fish include:

- — Growth rates that are 3-11 times faster with more efficient feed utilisation
- — Increased tolerance to cold water
- — Improved disease resistance.

Cloning

Another application of animal biotechnology is the use of somatic cell nuclear transfer to produce multiple copies of animals that are nearly identical copies of other animals (transgenic animals, genetically superior animals, or animals that produce high quantities of milk or have some other desirable trait, etc.). This process has been referred to as cloning. To date, somatic cell nuclear transfer has been used to clone cattle, sheep, pigs, goats, horses, mules, cats, rats, and mice. The technique involves culturing somatic cells from an appropriate tissue (fibroblasts) from the animal to be cloned. Nuclei from the cultured somatic cells are then microinjected into an enucleated oocyte obtained from another individual of the same or a closely related species.

Through a process that is not yet understood, the nucleus from the somatic cell is reprogrammed to a pattern of gene expression suitable for directing normal development of the embryo. After further culture and development in vitro, the embryos are transferred to a recipient female and ultimately will result in the birth of live offspring. The success rate for propagating animals by nuclear transfer is often less than 10 percent and depends on many factors, including the species, source of the recipient ova, cell type of the donor nuclei, treatment of donor cells prior to nuclear transfer, the techniques employed for nuclear transfer, etc.

Somatic cell nuclear transfer (SCNT) can be used in embryonic stem cell research, or in regenerative medicine where it is sometimes referred to "therapeutic cloning." It can also be used as the first step in the process of reproductive cloning. In SCNT the nucleus, which contains the organism's DNA, of a somatic cell (a body cell other than a sperm or egg cell) is removed and the rest of the cell is discarded. At the same time, the nucleus of an egg cell is removed. The nucleus of the somatic cell is then inserted into the enucleated egg cell. After being inserted into the egg, the somatic cell nucleus is reprogrammed by the host cell. The egg, now containing the nucleus of a somatic cell, is stimulated with an electric shock and the cell begins to divide. After many mitotic divisions in culture, this single cell forms a blastocyst with almost identical DNA to the original organism.

Some researchers use SCNT in stem cell research. The aim of carrying out this procedure is to obtain stem cells that are genetically matched to the donor organism. Presently, no human stem cell lines have been derived from SCNT research. A potential use of genetically-customised stem cells would be to create cell lines that have any genes linked to a particular disease. For example, if a person with Parkinson's disease donated his or her somatic cells, then the stem cells resulting SCNT would have any genes which contribute to Parkinson's disease. In this scenario, the disease-specific stem cell lines would be studied in order to better understand the disease.

In another scenario, genetically-customised stem cell lines would be generated for cell-based therapies to transplant to the patient. The resulting cells would be genetically identical to the somatic cell donor, thus avoiding any complications from immune system rejection. Only a handful of the labs in the world are currently using SCNT techniques in human stem cell research. In the United States, scientists at the Harvard University Stem Cell Institute, the University of California San Francisco, and possibly Advanced Cell Technology are currently researching a technique to use somatic cell nuclear transfer to produce embryonic stem cells. This technique is currently the basis for cloning animals and in theory could be used to clone humans. However, most researchers believe that in the foreseeable future it will not be possible to use this technique to produce a human clone that will develop to term.

Advances in biotechnology have allowed scientists to make genetically identical copies or clones of animals. Duplication of an organism's genome occurs naturally when identical twins are born or when a plant is grown from a cutting of another plant. However, the world really took notice of cloning in 1997 when a group of Scottish researchers announced the birth of Dolly the sheep, which had been cloned using a single cell from an adult sheep. Dolly had only one "parent;" her nuclear genome was exactly like her "mother's" instead of being a combination of two parents. Therefore, Dolly could generally be thought of as her mother's identical twin.

To produce Dolly, scientists took an egg from a sheep and removed its nucleus, rendering it unable to function or develop. Next, they took a cell with an intact genome from a different adult sheep (Dolly's "mother") and fused it to the sheep egg which lacked a genome. The egg, with its new genome, was stimulated to begin developing into an embryo and was implanted into a surrogate sheep where it grew normally, resulting in the birth of Dolly. Dolly later gave birth to normal lambs.

Researchers have cloned other mammals including cows, goats, pigs, and mice. However, the overall low rate of successful cloning and frequent occurrence of developmental abnormalities in cloned animals demonstrate the need for further research before cloning will be practical. It has also been reported that cloned animals may exhibit health problems throughout their life. Cloned animals may age prematurely as Dolly was diagnosed with arthritis at a seemingly young age and cloned mice had a shorter than normal life span. Additionally, it was demonstrated that cloned mice were both larger in size and heavier than a control group of non-cloned mice.

If advances in animal cloning technology were to overcome the current obstacles, the most obvious benefit would be the ability of a farmer to have a herd of superior performing animals in one generation. Breeding companies could sell cloned embryos in a manner similar to the way in which semen is currently marketed. A potential drawback of this practice would be the loss of genetic diversity in livestock herds, but this could be avoided by limiting the number of cloned embryos of a given animal that were sold.

It has also been proposed that cloning could be used to increase the population of animals in an endangered species. The mouflon sheep, which is a wild Mediterranean sheep with less than 1000 animals remaining, was successfully cloned. Additionally, scientists are attempting to clone an endangered wild Asian ox, called the guar (the first cloned guar died of an intestinal illness shortly after birth) and possibly the giant Panda. Although possible, a recovering population of cloned animals would be hindered by a lack of genetic diversity and would not address the larger issue of how the animal became endangered.

Issues Related to Somatic Cell Nuclear Transfer Technology

The generation of animals using nuclear transfer from somatic cells has received a great deal of recent attention, and it is clear that this technology is fast becoming a practical way to rapidly propagate animals with valuable properties. The DNA genomes of somatic cell nuclei used for this procedure differ in two important ways from those of germline cells. First, they have shortened telomeres at the ends of the chromosomes, a consequence of multiple rounds of cell division in the absence of telomerase, the enzyme responsible for a maintenance of telomere length. Since loss of a telomere length is the principal mechanism limiting the lifespan of cells in culture, the lack of appropriate length telomeres might be expected to reduce the lifespan of the newly generated offspring or their progeny, but, surprisingly, telomere length are restored to normal values following generation of cattle by somatic cell nuclear transfer, even when senescent cells are used to donate nuclei.

Second, the methylation state of the DNA of somatic cells is quite different from that of germline cells. Since methylation plays a major role in the overall regulation of gene expression, it might be expected that inappropriate methylation states might lead to gross developmental abnormalities and embryos produced by somatic cell nuclear transfer. Indeed, it is possible that the inability of the embryo to properly reprogramme methylation and expression is a major cause of the developmental abnormalities often seen in the generation of NT-produced embryos. However, the apparently rapid increase in success rate of this procedure with experience, combined with the fact that animals who survive to adulthood are apparently normal, implies that correct methylation can be restored in NT embryos under the proper conditions. "Correct conditions" might involve having the transferred nucleus and the proper stage of the cell cycle, but this point is controversial.

Furthermore in direct study, it was found that correct methylation and expression levels of several key genes were restored in piggy embryos derived from adult cell nuclei. Thus, although nuclear reprogramming is a significant practical issue in the efficient application of this technology, it does not appear to present as insurmountable a barrier as once thought. Apparently the developmental process has a much more robust error-correction system then believed possible a few years ago.

Midi carefully considered the possible concerns that might be raised by use of somatic cell nuclear transfer technology. A few issues regarding animal welfare could be identified, including the possibility of inappropriate gene expression during development due to altered methylation patterns, or other developmental problems, such as oversized fetuses, as well as concerns that the widespread application of such technology might reduce genetic diversity of animal population. However, the effects of cloning are more difficult to anticipate because competing processes are at issue. On the one hand, cloning by its nature produces identical copies of the particular individual, reducing genetic variability relative to what would have been transmitted via conventional breeding on the other hand, cloning makes it possible to save and utilise genetic variability that would not otherwise be available, for example, the genetic resources from paste deer proven to be high performing. The trade-off between the competing processes is hard to quantify in the absence of simulation modelling with validation from field observations. Whatever the mechanism causing it, loss of genetic diversity could limit the potential for future genetic improvements are breeds by selective breeding or biotechnological approaches. Furthermore, disease could spread through susceptible populations more rapidly than more genetically diverse populations.

This latter concern is well-documented and several studies illustrate the susceptibility of species with low genetic diversity to infectious disease. Diversity of animal populations, particularly at major histocompatibility (MHC) loci, is a major factor preventing spread of disease, particularly viral disease. Different MHC types recognise different viral or bacterial epitopes encoded by pathogens for presentation to the immune system. In genetically diverse populations, pathogens can evade the immune response only if they adapt to each individual MHC type following transmission from one individual to another. The requirement for this evolutionary process provides a population of animals with significant protection against the spread of infection. Pathogens can more easily evade post immune response in genetically uniform populations.

The consequences of the failure of immunorecognition is illustrated by the deadly epidemics of diseases—such as measles—spread by initial contact between Europeans and isolated New World populations that lack adequate MHC diversity. Not only could enhanced susceptibility creates significant risk for the spread of "new" infectious diseases in "monocultures" of cloned or highly inbred animal populations; it also could create new reservoirs for spread of zoonotic infections—like new strains of influenza—to humans. The seriousness of these concerns, particularly relative to current practice obviously must very considerably from one type of animal to another, and might be alleviated with further technological advances.

Gene Knockout Technology

Animal biotechnology also can knock out or inactivate a specific gene. A gene knockout is a genetically engineered organism that carries one or more genes in its chromosomes that have been made inoperative (have been "knocked out" of the organism). This is done for research purposes. Also known as knockout organisms or simply knockouts, they are used in learning about a gene that has been sequenced, but which has an unknown or incompletely known function. Researchers draw inferences from the difference between the knockout organism and normal individuals. The term also refers to the process of creating such an organism, as in "knocking out" a gene.

Knockout is accomplished through a combination of techniques, beginning in the test tube with a plasmid, a bacterial artificial chromosome or other DNA construct, and proceeding to cell culture. Individual cells are genetically transformed with the construct and—for knockouts in multicellular organisms—ultimately fused with a stem cell from a nascent embryo.

The construct is engineered to recombine with the target gene, which is accomplished by incorporating sequences from the gene itself into the construct. Recombination then occurs in the region of that sequence within the gene, resulting in the insertion of a foreign sequence to disrupt the gene. With its sequence interrupted, the altered gene in most cases will be translated into a nonfunctional protein, if it is translated at all.

A conditional knockout allows gene deletion in a tissue specific manner. Because recombination is a rare event in the case of most cells and most constructs, the foreign sequence chosen for insertion usually is a reporter. This enables easy selection of cells or individuals in which knockout was successful. In diploid organisms, which contain two alleles for most genes, and may as well contain several related genes that collaborate in the same role, additional rounds of transformation and selection are performed until every targeted gene is knocked out. Knock-in is similar to knock-out, but instead it replaces a gene with another instead of deleting it.

Knockout technology creates a possible source of replacement organs for humans. The process of transplanting cells, tissues, or organs from one species to another is referred to as “xenotransplantation.” Currently, the pig is the major animal being considered as a xenotransplant donor to humans. Unfortunately, pig cells and human cells are not immunologically compatible. Pig cells express a carbohydrate epitope (alpha1, 3 galactose) on their surface that is not normally found on human cells. Humans will generate antibodies to this epitope, which will result in acute rejection of the xenograft.

Genetic engineering is used to knock out or inactivate the pig gene (alpha1, 3 galactosyl transferase) that attaches this carbohydrate epitope on pig cells. Other examples of knockout technology in animals include inactivation of the prion-related peptide (PRP) gene that may generate animals resistant to diseases associated with prions (bovine spongiform encephalopathy [BSE], Creutzfeldt-Jakob Disease [CJD], scrapie, etc.).

Aquaculture Biotechnology

Use the tools of modern biotechnology to improve the health, reproduction, development, growth, and overall well-being of cultivated aquatic organisms; and promote the interdisciplinary development of environmentally sensitive, sustainable systems that will enable significant commercialisation of aquaculture. Aquaculture, which long has been practiced in Asia and is increasingly popular in the United States, Europe, and South America, will benefit tremendously from the use of new molecular tools and processes. With worldwide seafood demand projected to increase 70 percent in the next 35 years, and harvests from capture fisheries stable or declining, aquaculture will have to produce seven times as much seafood as it generates now to supply global demand by 2025. The use of modern biotechnology to intervene in the rearing process and enhance production of aquatic species holds great potential not only to meet this demand, but also to improve competitiveness in aquaculture.

The growth and international competitiveness of the aquaculture industry will be determined by the size of the resource investment in research and technology development. This investment should be made through a partnership of Federal and state agencies and the private sector. The Federal role is to provide leadership in supporting research to advance knowledge in important research areas and to facilitate the transfer of promising results and technologies to the private sector.

The major research issues in aquaculture are similar to those for other agricultural sectors, but the knowledge base for aquaculture is comparatively meager. Development of this knowledge is a particular challenge due to the diversity of cultured aquatic species and the systems for their production.

The application of biotechnology promises significant benefits to both producers and consumers of aquacultural products. The use of genetically enhanced organisms may improve production efficiency through improvements in growth rates, food conversion, disease resistance, and product quality and composition. The application of biotechnology to aquaculture also may help conserve wild species and genetic resources and provide unique models for biomedical research.

Aquaculture biotechnology can be applied to enhance reproduction and early development of cultivated aquatic organisms. The resulting benefits could include year-round production of gametes and fry of economically valuable species and creation of new markets for specialised, genetically improved broodstock. Similarly, biotechnology may provide techniques for improving the reproductive success and survival of endangered species, thereby helping to preserve the diversity of life on Earth. As a first step, research should be directed toward improving basic understanding of environmental, hormonal, biochemical, and genetic control of reproduction. More specifically, scientists must identify and understand the mechanisms of expression of genes involved in reproduction and development, improve technologies for cryo-preserving gametes and embryos, improve delivery systems for administration of natural and synthetic hormones, and enhance understanding of the pharmacokinetics of uptake and release of administered hormones.

Aquaculture biotechnology can be applied to improve the growth, development, and nutrition of cultivated aquatic organisms. Specifically, the tools of biotechnology can enhance understanding of the environmental, biochemical, hormonal, and genetic bases of growth and nutrition for all life stages. Biotechnology also can be used to expand knowledge about the absorption, digestion, and assimilation of nutrients and to increase the efficiency of these processes.

Aquaculture biotechnology offers substantial opportunities to improve the health and well-being of cultivated aquatic organisms. More than 50 diseases affect fish and shellfish cultured in the United States, causing losses of tens of millions of dollars annually. Biotechnology not only can improve the survival, growth, vigor, and well-being of cultivated stocks, but also can reduce disease transfer between cultivated and wild stocks. New products and market opportunities can be developed related to aquatic animal health and well-being.

The tools of molecular biology can provide a basic understanding of host immunity, resistance, and susceptibility to diseases and associated pathogens by furnishing information about life cycles and mechanisms of pathogenesis, antibiotic resistance, and disease transmission. Improved technologies must be developed for detecting and diagnosing pathogens and diseases and for enhancing the genetic basis of disease resistance, thereby reducing the need for antibiotics and other drugs.

Potential products resulting from this research include gene therapy techniques; broodstock free of pathogens; safe, effective prophylactic agents, including immune modulators, antigens, and vaccines; safe, effective therapeutic agents; and improved systems for administering prophylactic and therapeutic agents.

Aquaculture biotechnology can be employed to assess and improve the safety, freshness, color, flavor, texture, taste, nutritional characteristics, and shelf life of aquacultural food products. In addition, practical technologies can be developed to detect and assay toxins, contaminants, and residues in seafood, and to reduce or eliminate contaminants. There are also opportunities to apply biotechnology in improving seafood processing. Research should be conducted to develop and improve technologies for all these applications.

The preservation and enhancement of biodiversity in natural systems is an important Federal priority. Therefore, the Federal Government should encourage and support programs to maintain and enhance biodiversity in aquatic systems through cultivation and stocking of aquatic species. Biotechnology can be employed in two ways to conserve genetic resources of aquatic species. First, the tools of biotechnology can be used to identify and characterise important aquatic germplasm, including endangered species. Genomes of aquatic species can be analysed and characterised, and quantitative trait loci identified. Second, biotechnology can be applied to improve understanding of the molecular basis of gene regulation and expression as well as sex determination and thereby improve methods for defining species, stocks, and populations. Approaches include developing marker- assisted selection technologies, improving precision and efficiency of transgenic techniques, and improving technologies for the cryo-preservation of gametes and embryos. Ultimately, stocking certain areas with selected, cultivated species and strains could help maintain biodiversity in natural aquatic ecosystems.

Techniques of Animal Genetic Engineering

The art and science of producing genetically engineered animals have advanced there are rapidly in the past few years. It now is possible to generate animals with useful novel properties for dairy, meat, or fibber production, for environmental control of waste production, that produce useful products for biomedical purposes or other human consumption, or that are nearly identical copies of animals chosen for useful traits, such as milk or meat production, pie fertility, and the like. There are a number of methods presently employed for genetic engineering of various animal species most of these were developed originally in mouse and Drosophila models and have only more recently been extended to other domesticated animals. Access to the germline of mammals can be obtained by:

(1) direct manipulation of the fertilised egg, followed by its implantation into the uterus;

(2) manipulation of the sperm used to generate the zygote;

(3) manipulation of early embryonic tissue in place;

(4) the use of embryonic stem (ES) cell lines which, after manipulation and selection ex vivo,a nd then be introduced into early embryos, some of whose germline will develop from the ES cells; and

(5) phase manipulation of cultured somatic cells, whose nuclei than can be transferred into enucleated oocytes and thereby provide the genetic information required to produce a whole animal.

The last two methods have the advantage of allowing cells containing the modification of interest to be selected prior to undertaking the expensive and lengthy process of generating animals. Usable ES cells are not available for all species of interest, however, and generation of embryos by nuclear transfer (NT) from somatic cells is becoming the method of choice for genetic engineering and duplication of nearly genetically identical animals.

Manipulation of the avian germline is difficult, since ES lines are not available in the early embryos difficult to access. Much current work focuses on the use of blastodermal cells or primordial germ cells, which can be cultured briefly and manipulated to modify the germline, prior to introduction into fresh embryos to create chimeras from which modified lines can eventually be developed, albeit with low efficiency phase. There are two basic approaches presently in use for inserting DNA into vertebrate germline cells, transfection and infection with retrovirus vectors. A third approach, based on the use of mobile genetic elements, has been commonly used for insects, and is being explored for germline modification of vertebrates.

Transfection

Transfection methods include:

(1) direct microinjection of DNA into the cell nucleus;

(2) electroporation—introduction of DNA through transient pores created by controlled electrical pulses;

(3) use of polycations to neutralise charges on DNA and cell surface that prevent efficient uptake of DNA;

(4) lipofection, or enclosure of DNA in lipid vesicles that enter a cell by membrane infusion much in the manner of a virus, and

(5) sperm-mediated transfection, possibly in conjunction with intracytoplasmic sperm injection (ICSI) or electroporation.

The manner of introduction of DNA is a technical issue, determined empirically for each system, and makes little difference to the final outcome. In general, the structure of DNA introduced into a cell by any of these methods is highly variable and uncertain. Often, only a fragment of the transfected DNA is integrated into the chromosome, frequently in multiple copies, often in long tandem arras. When transfecting cultured somatic or ES cells, of selectable marker, such as the gene encoding phosphotransferase, is often included as part of the DNA to permit selection for its presence either in eukaryotic cell lines or in the bacteria in which the DNA was mass-produced.

Retrovirus Vectors

Retroviruses are infectious elements that replicate by a unique process involving copying of the viral RNA genome into DNA followed by its specific and stable introduction into host cell DNA (integration). The integrated DNA than can be expressed using the normal transcriptional machinery of the cell. Retroviruses commonly are used to introduce genes of interest into cells and culture or into somatic tissue in experimental animals. They also have been used for germline modification of fish, mollusks, chickens, mice, and more recently, cattle.

To make a retrovirus vector, a DNA construct containing the gene of interest is flanked by sequences necessary for replication as a virus. The sequences include transcriptional promoters in the long terminal repeats, which flanked the integrated DNA, or provirus. Signals necessary for packaging of the transcript in virions (virus particles), for reverse transcription, and for integration of the resulting DNA also must be included. Introduction of such a DNA construct into cells that express viral proteins, but that are incapable of making infectious virus leads to the creation of infectious virions containing an RNA copy of the gene of interest. After infection of cells with such virions, the RNA is copied into DNA and integrated at random sites in the cell genome. Again, selectable markers often are included in the construct to select cells containing the desired virus construct.

Transposons

Transposons are DNA elements that can transfer their information from one site to another in the same cell. A variety of transposons have been found in insects, and fish, and some are routinely used as vectors for the generation of transgenic insects. No active transposons of these types have been observed in mammals, although human genome contains thousands of copies of the DNA sequence related to the mariner transposon of Drosophila phase, suggesting that there might have been active elements in our recent evolutionary history. Nevertheless, several recent reports suggest that naturally-occurring transposons found in insects, or even bacteria, might provide useful and

Efficient means for introducing genes into the germline of animals. Mariner, for example, has been shown to be active in chick zygotes, transferring its DNA from 8 microinjection plasmid into the germline, albeit at low efficiency. A modified version of Sleeping Beauty, a related elements from fish, has been developed to give a high yield of germline or somatic transformants in cultured cells and laboratory mice. In practice, it remains to be seen whether — and how efficiently — genes of interest can be transferred in this way. Another transposon system being investigated for this purpose is the T DNA of Agrobacterium tumefaciens, a natural pathogen of plants. The bacteria and can't use of plants cells, leading to transposases and of the DNA in the nuclear DNA of the host. This technique is widely used for the generation of transgenic plants. Remarkably, it recently has been shown that Agrobacterium can't use with human cells and culture, leading to transfer of T DNA carrying a marker gene. None of the transposons-based techniques currently are used for the generation of transgenic livestock, but they might lead to more efficient methods for this purpose.

Directed Genetic Manipulation

Another goal of transgenic technology is the creation of engineered animals thatne that has been engineered any specific way. For example, transplantation of organs are tissues from non-primates (such as pigs) to humans (xenotransplantation) is currently impossible, in part due to a dramatic immune response against a cell-surface carbohydrate not found in old-world primates. Inactivation of the enzyme from donor pigs could alleviate this problem, and pigs with one allele of the gene encoding this enzyme recently have been produced, giving rise to expectation that completely GT-deficient animals soon will be available. Another important goal is the elimination from cattle of the gene encoding prion-related protein (PrP), the protein associated with scrapie and sheep and bovine spongiform encephalopathy (BSE or mad cow disease). Removal of this gene from mice has, at most, subtle phenotypic consequences, yet renders them completely resistant to these diseases. If the mouse model holds true in cattle, homozygous knockout of bovine PrP could lead to the elimination of BSE.

Gene knockout/knockin technology is well-established as an experimental tool in mice, due to the availability of ES cell lines. The principle is to take advantage of a rather rare event that occurs after introduction of DNA into cells-homologous recombination between identical sequences in the genome and the transfecting DNA. In the most common protocol, a selectable marker is inserted within a piece of DNA corresponding to the portion of a gene of interest. After transfection of cells by this construct and selection for the marker, the selected cells are screened to identify the small fraction that has one copy of the gene of interest disrupted by the marker. Progeny animals derived from the cells will be at heterozygous for the "knocked-out" for "knocked-in" gene; breeding to obtain homozygous animals is straightforward. Because the process is so inefficient, very large numbers of transfected cells was be screened, making the use of cultured cells essential, since it would be impracticable to screen large numbers of progeny from microinjected eggs. The galactosyl transferase-knockout pigs discussed above were generated from cultured fetal fibroblasts manipulated in this way.

Propagation by Nuclear Transfer

In February of 1997, Dolly the sheep was introduced and the public subsequently was inundated with opinions about the power and potential of creating new animals from somatic cells. Dolly represented the most recent advance and genetic technology — the production of multiple individuals nearly genetically identical to an adult animal. In these technique, somatic cells from the appropriate tissue are grown in culture, in their nuclei are injected into enucleated oocytes obtained from another individual of the same or closely related species. After a further. Of culture, the partially developed embryos are transformed into a foster mother. This technology is being developed rapidly for many species of interest, and promises to become a rapid and efficient means of propagating domestic animals with desired traits, whether those are naturally derived and selected or genetically engineered. The process often is referred to as "cloning". The nuclear transfer technique was based on previous studies and frogs conducted during the previous five decades, but, until Dolly, it was unclear whether nuclei from highly differentiated somatic cells could be reprogrammed to a pattern of gene expression suitable for directing normal development of a mammalian embryo.

The ability to reprogrammed the nucleus of donor cells for successful nuclear transfer appeared initially to require the use of methods that facilitates cell cycle synchrony, since only after the donor cells were induced to become quiescent could offspring be obtained by transfer of the donor nucleus to enucleated oocytes. The necessity for quiescence is not as clear today, since nuclei from actively dividing cells have now also been used successfully for this purpose. The ability to reprogrammed the donor cells also depends on the species and nuclear transfer procedure. Currently, only oocytes can be used successfully, as they are the only recipients cells that convert differentiated nuclei into undifferentiated stages resembling pronuclei in freshly fertilised zygotes, a step which is essential for the completed development of reconstructed embryo. How the enucleated oocyte accomplishes this reprogramming is currently unknown.

At present, propagation of animals by nuclear transfer is inefficient, with an average of less than 10 percent of the embryos resulting in live offspring, although the success rate appears to be increasing with experience. Most of the failures occur during development, and there appears to be increased rate of perinatal death relative to normally-conceived offspring. In cattle, at least, the developmental and perinatal problems appear to be as much of function of the in vitro culture technology fan of the nuclear transfer itself. However, even with this existing low efficiency, there are many potential applications for reproducing highly desired genotypes, including rare or endangered species, household pets, delete sires or dams, breeds with desirable production traits but low fertility, sterile animals such as castrates a and mules, or transgenic animals that have high value, and for which rapid propagation is desirable.

In other important to application of this technology is in the dissemination of germ plasm as embryos and consequent reduction of the associated risk of disease spread. Is also important to note that there are significant differences between cattle & in terms of the utility of this technique. In cattle, ejaculate from a single bull can be used to breed 4 to 500 females in AI programs. In contrast, the ejaculate from a single bull or can be used to breed only 10 to 20 females. Thus embryos obtained from NT might be the method of choice for the dissemination of swine germplasm rather than AI. A dissemination of swine germplasm rather than AI. A number of variables influence the success rate of nuclear transfer. These include: species, source of the recipient ova, cell type of donor nuclei, treatment of donor cells prior to nuclear transfer, and the techniques employed for nuclear transfer.

Issues of Germline Technology

A key consideration in development of transgenic animals is ensuring that the gene product of interest is expressed in the current tissue, and at the appropriate level and time. Specifics of how this goal is accomplished very considerably from one system to another, but some general principles can be elucidated. First, the natural regulatory elements (promoters) for most genes that direct tissue-specific transcription are complex, large, and poorly understood. For this reason, well-characterised promoters for other genes are appended to DNA encoding the desired gene product. Second, the expression of Trans genes, especially those derived by transfection, is strongly under the influence of control elements in the DNA around the integration site. These positional effects often lead to silencing of the gene of interest, or, more rarely, to unregulated expression. They can be alleviated to some extent by the inclusion of sequences, such as insulators, or locus-control regions, but it is impossible to predict whether a given construct will show the desired pattern of expression after integration. While these effects do not directly affect the safety or utility of those animals that are eventually used, they do introduce considerable inefficiency into the system.

A further problem with obtaining correct expression of and introduced Trans gene is that introduced genes are subjected to silencing by processes including methylation of C residues at CpG dinucleotides, which frequently are found in chromosomal regions important to the regulation of gene expression. Methylation is a major mechanism for turning off the expression of inappropriate genes and somatic cells. Silencing can occur in somatic tissues, but is particularly acute with introduced genes after passage through the germline, where there is widespread DNA methylation at an early stage of embryogenesis.

Normal genes in a proper place have signals — in most cases unknown — that reverse the methylation at the appropriate developmental stage. Such signals generally are not present on many commonly used promoter elements, such as retroviral LTR sequences, and expression directed by these elements really survives passage through the germline. As a natural example, humans carry thousands of endogenous proviruses that have resulted from retroviral infection of the germline of our distant ancestors; yet only a few are ever expressed to any significant level.

Proviruses based on commonly-used murine leukemia virus (MLV) factors introduced by deliberate infection of the germline suffer a similar fate. Recently, it has been found that factors based on human immunodeficiency virus can be used to efficiently insert genes into the germline of mice, and that genes inserted in this way are not subject to silencing following germline transmission. Such factors promised to yield more efficient and reliable means of duration of obtaining transgenic animals of many species. Similar vector systems based on the distant related feline immunodeficiency virus (FIV) and bovine immunodeficiency virus (BIV) are also being developed. Again, methylation-induced shutoff of gene expression is an issue affecting the strategy and efficiency of production of transgenic animals, much less their safety as producers of useful products.

Germline on occasion remains a hit-or-mess technology, and with most techniques only a very small fraction of the progeny obtained as the desired properties of expression, copy number, and lack of genetic damage. Dust, large numbers of animals must be screened for the presence and copy number of the inserted sequence, for its properly regulated expression, for the ability of this expression to some 5 transmission through the germline and, finally for the desired phenotypic characteristics and absence of unintended genetic side effects. Such screening can require several generations of breeding before one can be confident of and absence of assessed of genetic damage, and the failure rate of the overall process is very high.

As nuclear transfer technology improves, techniques requiring direct introduction of DNA into the animal germline followed by extensive screening of progeny are likely to be replaced by much simpler manipulation and selection of cells and culture, followed by recreation of animals with the desired properties directly from the nuclei of the manipulated cells. There are a number of safety issues that arise as a consequence of manipulation of the germline. These can be divided into several levels of concern: to the animal to the human handler, recipient, or the user of the animal or its products; to the human population as a whole; and to the environment.

Genetic Side Effects

Introduction of DNA into random sites in the germline is a mutagenic event that will affect any gene that happens to be at or near the site of integration. The most obvious effect is the disruption of the integrity of the gene into which the insertion occurs. Since a large fraction of the mammalian genome is noncoding DNA derived from various kinds of silenced transposable elements, not all integration advance will lead to gene inactivation; however a fraction of animals selected for the presence of the transgene has been found to carry associated genetic lesions. In mice, for example, it has been estimated at about 5 percent of the MLV proviruses integrated into the germline have led to mutations of this sort. Direct DNA introduction can lead to numbers of integrated copies at multiple sites, leading to a risk of creating animals with a variety of genetic defects, which should be carefully screened for in the course of subsequent breeding.

For example, one of the very first transgenic mouse lines generated, intended to contain an inserted active oncogene, also suffered a lesion that caused a severe recessive developmental limb defect. A number of other examples of insertional inactivation by Trans gene's introduced into mice are known, and this approach has been proposed as a useful technique for mutagenisis. Two additional points should be noted. First, the successful transfected embryo They have inserted DNA sequences in addition to those that expressed the transgene, so the point of damage can be at a location different from the active transgene. Second damage of this sort is often (but not always) recessive, so that it can only be detected by inbreeding to derive animals homozygous at the site(s) of the inserted DNA, adding to the complexity of the screening process.

A related effect is the activation of expression of genes in the vicinity of the transgene through the action of the introduced promoter elements. This sort of inappropriate activation of expression is the mechanism of cancer induction in animals infected by a variety of retroviruses, and it has been well studied as a model for oncogenesis. There are a number of mechanisms by which the expression of genes adjacent to the integration site can be activated, including promoter and enhancer insertion, as well as gene fusion and introduction of elements that stabilise messenger RNA. Indeed, alteration of expression of genes at genome sites far removed from a transgene has been reported in cell lines, apparently do to altered methylation. Whether this effect also occurs in transgenic animals is not known. Activation effects are likely to reveal themselves as dominant mutations that can have a variety of phenotypic consequences, from derailing normal development to causing a high rate of cancer later in life.

Effects of Modification

The genetic engineering itself can often have unexpected effects on the physiology of the engineered organisms. One example of such an unwanted effect relates to the xenotransplantation model described above. The galactosyl transferase deficiency in humans, which leads to hyperacute rejection of organ from pigs, is also thought to offer a level of protection against zoonotic infection by enveloped viruses. This effect occurs because the surface proteins of viruses produced by nonhuman cells are also engineered with the same galactosyl-galactose structure found on post cell proteins, and are therefore subject to the same potent immune response. This response would lead to the rapid elimination of viruses transmitted from animals before infection could occur. Pigs that are engineered by knockout of this gene would, therefore, have the potential to transmit viruses, such as influenza, much more readily to human handler's.

A related concern is that human cell-surface proteins introduced into animal species as transgenes could render those animals susceptible to human viruses, increasing their risk of disease and providing alternative hosts for the spread of human disease. For example, the human poliovirus receptor (CD155) renders mice susceptible to poliovirus infection when introduced as a transgene. Also, the human complement-response modifying proteins CD46 and CD55, which are being introduced into pigs to protect xenografts from rejection, also serve as receptors for human viruses — measles and Coxsackie, respectively. Their presence in transgenic pigs not only could render these animals susceptible to infection by the human viruses, but could also provide a new evolutionary pathway for adaptation of pig irises the human cells. Since the receptors for many other viruses have not yet been identified, the potential for this sort of surprise exists whenever a human cell-surface proteins introduced into an other animal species.

Marker Genes

Factor constructs used for creating transgenic organisms usually contain genes other than the desired transgene. These genes are typically drug resistance markers obtained from bacteria, which also can confer resistance to the same or similar drugs on eukaryotic cells. The neo gene and coating neomycin phosphotransferase, for example, has been used widely for selecting cells and culture infected with retroviruses or other gene constructs (vectors). In most cases, marker genes remain in vectors used for generation of transgenic (especially knockout) animals. While many researchers in the field consider them a relatively harmless convenience, there is a potential for them to cause on desired side effects to the host species (such as aiding in the generation of novel antibiotic resistant pathogens) or the ultimate consumer (such as acting as novel allergens). While their potential for real harm is probably very small, it is difficult to impossible to prove that marker genes are harmless in consumer products. Such genes are usually unnecessary to the product itself and can usually be dispensed with by sound experimental design. Their presence raises concerns for food as well as environmental safety of the genetically engineered animal product.

Undesired Inserts

In addition to insertion of the correct element at multiple locations, the preparation of material used to generate the transgene (or knockout) might contain additional sequences unrelated (or partially related) to the one of interest and intent. Even extensively purified DNA fragments derive from plasmids grown in E. coli might still contain large amounts of contaminating the tear real derived from the host bacterium. Because such fragments can be heterogeneous in size and sequence, there difficult to detect in DNA preparations by standard methods like gel electrophoresis.

A particular problem in this regard arises with retroviral vectors, because post cells especially of mouse origin often contain large numbers of endogenous virus and virus-like sequences that can, in some cases, constitute a majority of the genomes present in vector preparations. Inadvertent introduction of such sequences into the germline of transgenic animals not only has a potential for creating unintended genetic damage, but also can contribute by recombination to the generation of novel infectious viruses. A well-known example is the inadvertent generation of replication-competent MLV's containing multiple such recombinants during the growth of a vector containing a globin gene. These viruses were highly pathogens in a rhesus monkeys, causing a fatal lymphoma, similar to the disease induced by MLV in mice.

Potential for Mobilisation

When viral vectors are used for the introduction of genes into the germline of animals, there exists a potential for inadvertent transmission of the gene to other animals. This undesirable effect could occur if such an animal was infected with the virus sufficiently similar to the vector to package the vector into virions. For example, if a transgenic chicken or created using an avian retrovirus vector, then infection of the transgenic animals with any related virus could lead to the production and release all virus that could transmit the gene to other animals where its presence and expression might be highly undesirable.

Generation of replicating virus could occur and the absence of exogenous infection, since many species contain endogenous retroviruses in their genomes that could serve as agents of this kind of mobilisation. For example, and cats caring murine leukemia virus (MLV)-based factor constructs, the introduced genes could be mobilised to other cats by the endogenous feline leukemia viruses found in most animals. The use of vectors based on HIV has the potential to improve the efficiency of introduction of new genes into the germline of many animal species. Such germline vectors could, in principle, also be mobilised by HIV or a sufficiently close relative. Viruses closely related to HIV are found Olean African primates; however, viruses of the same genus (Lentivirus) are fairly common and cats, cattle (bovine immunodeficiency virus or BIV), and sheep. Despite the distant relationship, FIV has been shown to transfer HIV-based vector constructs from oneself to another, raising a serious concern that similar transfer of genes introduced by an HIV factor could be mobilised among animals infected with these common viruses.

A related concern arises with the use of mariner and related transposons (including sleeping beauty) to introduce germline DNA. Related elements have been found in large numbers in the human genome and planaria, nematodes, centipedes, mites, insects, and humans, suggesting the possibility of horisontal gene flow via transposons in among highly diverse hosts. The possible importance of gene flow in eukaryotes is the ubiquity of transposable elements and endogenous retroviruses in the genomic DNA, with no known means for their distribution other than by horisontal gene transfer. It should be noted that any groups using transposable elements for genetic engineering could express the transposase or hopase in the trans configuration and then delete the gene from these enzymes from the transgene constructs, so that once inserted into the host's chromosome, the element is a mobilised. Were of this a requirement applied to transposable element vector systems for genetic engineering of animals, the hazards at issue could be minimised or eliminated, so long as active elements capable of mobilised in the introduced sequences were not already present in the host animal.

Creation of New Pathogens

In addition to their potential for mobilisation by interaction with related viruses, transgene sequences also can contribute elements to infecting agents that might modify their ability to cause disease. The donation of drug resistance genes to bacteria as a consequence of their widespread presence and transgenic livestock, is one theoretical example, although the resistance gene would have to be one not found in the environment for the risk of such an event to be significantly enhanced over the natural background. Another example is the possible generation of new retroviruses following recombination between endogenous or exogenous viruses and ones used as vectors for transgenes. This recombination event could result in the provision of new genes or regulatory elements that could adversely modify the pathogenic potential of the infecting virus.

A recent, natural example is the generation, through recombination between infectious avian retrovirus and a distantly related endogenous element among of a highly in virulent virus, called HPRS-103, or subgroup J Avian Leukemia Virus. This virus apparently rose as the result of a single, very rare event, but subsequently has been spread worldwide, and has become a source of considerable economic loss to poultry breeders.

Policy Concerns

Animal agriculture is being transformed by rapid advances in biotechnology — a term often used as a synonym for a variety of technologies that includes genetic engineering, genetic modification, transgenics, recombinant DNA techniques, and cloning, among others. Producers are interested in the application of biotechnology to improve productivity, consistency, and quality; to introduce new food, fiber, and medical products; and to protect the environment. Potential human health applications of transgenic animals include producing biopharmaceuticals and generating organs, tissues, and cells for xenotransplantation. Criticisms of such applications range from food safety and social resistance to potential negative impacts on animal welfare and on ecosystems. Questions also have arisen about the adequacy of the current regulatory structure to manage the complex interaction among environmental issues, public health, and food safety that developments in animal biotechnology may create.

Environmental Issues

Environmental concerns arising from emerging animal biotechnologies are largely speculative at this time because few products have been commercialised. Industrial developers of agricultural biotechnology might argue that more efficient production of animal-based feeds could reduce the resources necessary to produce food and, thereby, reduce the environmental burden of animal production. Should the development and widespread adoption of the "EnviroPig", which produces less phosphorus in its waste, occur, it might be considered by some to be a positive environmental benefit of agricultural biotechnology.

Many scientific studies reportes the potential negative environmental impacts of genetically altered animals. Escape, survival, and gene flow into wild populations are identified as major concerns. Of most concern was the escape into the environment of "super" salmon that have been genetically modified for rapid growth, and the likelihood that they could then breed with wild populations in the environment. Other genetically altered animals such as fish, insects, and shellfish could also potentially escape into natural environments and become feral, disrupt ecosystems, or introduce novel genes in a natural population.

Food Safety

Unexpected and unintended compositional changes arise with all forms of plant and animal genetic modification, including GE. So far, no GE-related adverse human health effects have been documented. However, there are "sizeable gaps" in the ability to identify compositional changes caused by all forms of genetic modification — whether GE or conventional — and their relevance for human health, and they recommended new approaches for assessing the safety of new foods both before and after they enter the market. Previous research and experience with commercialising transgenic plants suggest that negative effects on human health are virtually nonexistent. While not asserting that genetically modified organisms necessarily generate health problems, more recently reported research in peer-reviewed scientific journals has suggested that GMOs may raise food safety concerns:

The scientific principles for assessing the safety of GE animals are "qualitatively the same" as for non-GE animals.However, because GE can introduce new proteins into foods, the potential for allergenicity, bioactivity, and/or toxicity responses should be considered, they said. Animals genetically engineered for nonfood products like pharmaceuticals or replacement organs might be of concern if such animals entered or affected the food supply.Cloning. The biotechnology industry so far has agreed not to market milk and meat from cloned animals while it awaits risk assessment on the safety of such products.

Embryonic splitting and nuclear transfer using embryonic cells were used with some dairy cows to successfully produce genetically valuable offspring that were milked commercially and whose milk and meat entered the food supply. Few concerns were raised by the report about using these types of cloned animals for food, and they are believed to pose a low level of food safety concern. However, evaluating clonedanimal food composition would be prudent to minimise any food safety concerns. The products of offspring of cloned animals were regarded as posing no food safety concern because they are the result of natural matings.Other issues, notably consumer acceptance, social values, and animal welfare could eventually overshadow any lingering questions about human health.

Consumer and Social Acceptance

Criteria for selecting desirable traits to be produced through transgenic animals will likely be based on the demand for specific commercial characteristics. Even if scientific evidence is convincing that GE and cloned animal products are safe and beneficial for human consumption or economically valuable to producers, other concerns may limit marketplace and consumer acceptance. Polls in recent years in the United States indicate that public knowledge about food and biotechnology generally remains limited. In two 2005 surveys, approximately half of those surveyed expressed opposition to the use of biotechnology in the food supply.More than half of those in the Pew-sponsored poll said they opposed research into genetically modified animals, although opposition declined with increased knowledge. Many Americans have heard about animal cloning; two-thirds expressed discomfort with it — more of them out of religious or ethical concerns than food safety concerns. A majority of respondents to the Pew survey believe that regulators should take into account ethical and moral considerations.

Consumers may be less willing to accept the practice of genetically modifying animals than plants, some have argued, observing that people relate differently to animals, which many recognise as sentient beings. Some observers have expressed the concern that cloning farm animals might lead more quickly to human and pet cloning, which those observers oppose. Others believe that modifying animals, for example, to save human lives through xenotransplantation or the production of some important drug, might be more acceptable than doing so simply to produce more or cheaper food. Further, science alone cannot resolve ethical views which appear to vary widely:

Some people, irrespective of the application of technology, consider genetic engineering of animals fundamentally unethical. Others, however, hold that the ethical significance of animal biotechnologies must derive from the risk and benefits to people, the animals, and/or the environment. Yet another view focuses on the right of humans to know what they are eating or how their food or pharmaceuticals are being produced and therefore labeling becomes an issue to be addressed.

Regulatory decisions and enforcement involving animal biotechnology "are difficult in the absence of an ethical framework."Some believe that segregating and labeling the products of biotechnology in agriculture, including meat and milk, would enable the consumer to choose whether or not to buy such products. Both failing to segregate and label biotechnology products as well as doing so can contribute to public suspicion that these products are flawed or different in some negative way, which may lead to contradictory policy decisions.Others counter that such products essentially are the same — and subjected to the same rigorous safety standards — as more conventional products, and therefore should be treated no differently in the marketplace.

Animal Welfare

Some aspects of gene transfer have the potential to create infectious disease hazards or impaired reproduction. Looming large in the ethical debate are questions about whether genetic modifications, cloning, and other technologies, stress animals unnecessarily, subject them to higher rates of disease and injury, and hasten death. For example, ruminants produced by in vitro culture or nuclear cell transfer methods tend to have higher birth weights and longer gestation periods than those produced by artificial insemination, creating potential calving problems. Nuclear transfer techniques to propagate genetic modifications may increase risks to the reproductive health and welfare of both the surrogate female animals and their transgenic offspring. Reserach studies have cited evidence of problems such as anatomical, physiological, or behavioral abnormalities in many transgenic animals. Some scientists have countered that animal welfare problems have been exaggerated, particularly as the technologies have advanced. Most appear to agree, however, that animals originating from some forms of genetic modification may require closer observation and care.

References

Bruce D. and A. Bruce, 1999. *Engineering Genesis: The Ethics of Genetic Engineering*, London: Earthscan Publications.

Friedman, Yali. 2008. *Building Biotechnology: Starting, Managing, and Understanding Biotechnology Companies*. Washington, DC: Logos Press.

Oliver, Richard W. *The Coming Biotech Age*.

Thieman, W.J.; Palladino, M.A. 2008. *Introduction to Biotechnology*. Pearson/Benjamin Cummings

3

Genetic Modification of Fish

In developing more effective and sustainable exploitation of fish populations, the systematic use of the considerable battery of genetical techniques now available is still relatively underdeveloped. This statement holds whether we consider natural populations, enhanced populations or cultured stocks. However, there is increasing recognition that combining well established techniques such as the selective breeding programme carried out on Atlantic salmon with appropriate molecular techniques, should yield valuable results in aquaculture.

Of the range of molecular techniques available, some may be considered as "platform technologies" following the terminology of Hew and Fletcher, and of these it seems likely that transgenesis will be one of the most significant.

The production of appropriate genetically modified organisms or GMOs (in some cases combined with other forms of genetic improvement) offers considerable opportunities for more efficient and more effective aquaculture across a wide range of species.

The value of GMOs in agronomy is already widely accepted, as the area sown to transgenic crop species world wide exceeds 60 million hectares and this area is increasing rapidly year by year. However, both in terrestrial and aquatic animal species, while many GMOs have been produced, we have not succeeded in obtaining any hard evidence of commercial use. In aquaculture Dunham has a statement that this is taking place in New Zealand and Scotland though we have no further evidence that this is indeed the case.

Carr refers to small scale production in Cuba though several Cuban scientists have assured one of us that there is no commercial production in that country. However, given the drive towards large increases in aquaculture production evident in some countries, e.g. China it seems inevitable that commercial production of aquatic GMOs will not be long in coming.

NATURE OF GMOS

There are some difficulties in discussing GMOs because of the different definitions employed. At a world level the most recent, and probably the most useful, pronouncement is the so-called Cartagena Protocol on Biosafety to the 1992 Convention on Biological Diversity.

The Protocol does not refer to genetically modified organisms but rather, for reasons that are not explicit, to "living modified organisms" but it is clear that the two terms should be regarded as synonymous.

In the protocol the following definitions are used:

— "Living modified organism" means any living organism that possesses a novel combination of genetic material obtained through the use of modern biotechnology;

— "Living organism" means any biological entity capable of transferring or replicating genetic material, including sterile organisms, viruses and viroids;

— "Modern biotechnology" means the application of:

  a. *In vitro* nucleic acid techniques, including recombinant deoxyribonucleic acid (DNA) and direct injection of nucleic acid into cells or organelles, or

  b. Fusion of cells beyond the taxonomic family, that overcome natural physiological reproductive or recombination barriers and that are not techniques used in traditional breeding and selection;

This set of definitions is reasonably rational and scientifically defensible as it applies to transgenics but it does not specifically include nuclear transplants and nuclear cloning (not normally understood as cell fusion) and "fusion beyond the taxonomic family" is a somewhat arbitrary boundary not further justified.

For the purposes of this chapter, we restrict our considerations to transgenic organisms although we note in passing the interesting potential in aquaculture for nuclear cloning.

Transgenics are those organisms into which (donor) DNA has been inserted and incorporated using the *in vitro* techniques of genetic engineering. It is important to recognise that, as noted in the Cartagena protocol, it is the technology which defines a GMO not the source of the donor DNA and the reference frequently seen to "foreign" DNA is often misplaced. Thus a carp receiving and incorporating within its genome a sequence from its own DNA, is equally a transgenic, with the same fish receiving and incorporating a sequence from a daffodil. We can, however, usefully distinguish here between autotransgenic and allotransgenic, especially as this constitutes a particularly important factor when considering aquatic species. However, we may note that there is some scientific opinion which seeks to redefine transgenics as referring only to the allotransgenic category referred to above.

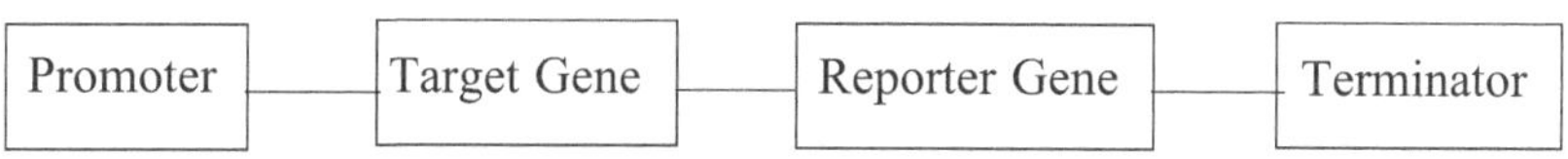

Figure 1. Diagram of DNA sequence of a typical construct for making transgenic organisms.

Why are GMOs Produced?

The main reasons for genetic manipulation of species used in aquaculture are all directly connected to improved output/input ratios. They are:

a) to enhance growth and/or efficiency of food conversion,

b) to enhance commercially significant flesh characteristics,

c) to control reproductive activity and/or sexual phenotype,

d) to increase resistance of species to pathogens/parasites,

e) to increase tolerance to/of environmental variables such as temperature,

f) to modify behaviour, e.g. aggression, and

g) to control fertility and/or viability.

While all of these targets are desirable in aquaculture, work up to now has been focused primarily upon points a. And e.

GMOS in Aquatic Species

The first transgenic animal to be produced was a mouse. The first recorded instances of production of transgenics in aquatic species are those of Maclean and Talwar in rainbow trout and Zhu *et al.* in goldfish. Since then many species have been used to produce GMOs as shown in Table 1. The list represents an amalgam of species significant in aquaculture with species amenable to laboratory culture and with short life cycles used particularly for studies of gene action, studies which of course form the platform for better understanding and hence better production in aquaculture.

Table 1. Aquatic species in which GMOs have been induced

Common name	*Latin name*	*Number of constructs employed to generate transgenics*
Fish		
Atlantic salmon	*Salmo salar*	6
Coho salmon	*Oncorhynchus kisutch*	4
Chinook salmon	*Oncorhynchus tshawytscha*	3
Tilapia	*Oreochromis spp.*	12
Medaka	*Oryzias latipes*	17
Zebra fish	*Brachydanio rerio*	14
Common carp	*Cyprinus carpio*	14
Channel catfish	*Ictalurus punctatus*	9
African catfish	*Clarias gariepinus*	1
Rainbow trout	*Oncorhyncus mykiss*	7
Cutthroat trout	*Oncorhyncus clarki*	1
Goldfish	*Carrassius auratus*	5
Northern pike	*Esox lucius*	2
Loach	*Misgurnus anguillicaudatus*	2
Sea bream	*Sparus aurata*	2
Red Sea Bream	*Pagrus major*	1
Blunt snout bream	*Megalobrama amblycephala*	1
Nigorobuna	*Carrassius auratus grandoculis*	1
Walleye	*Stizostedion vitreum*	1
Others		
Brine shrimp	*Artemia franciscana*	1
Seaweed	*Laminaria japonica*	1
	Undaria pinnatifida	
Sea Urchin	*Strongylocentrotus purpuratus*	1
	Paracentrotus lividus	
	Arbacia lixula	
Abalone	*Haliotus rufescens*	1

PROCESS OF GENETIC MODIFICATION

Production of GMOs is a multistage process which can be summarized as follows:

1. identification of the gene interest;
2. isolation of the gene of interest;
3. amplifying the gene to produce many copies;
4. associating the gene with an appropriate promoter and poly A sequence and insertion into plasmids;
5. multiplying the plasmid in bacteria and recovering the cloned construct for injection;
6. transference of the construct into the recipient tissue, usually fertilized eggs;
7. integration of gene into recipient genome;
8. expression of gene in recipient genome; and
9. inheritance of gene through further generations.

Choice of Target Genes

As shown in Table 2 the most popular gene used in aquatic species is growth hormone (GH) for reasons that are obvious. GH has been widely used in terrestrial species and as the gene sequence is highly conserved; the product is readily utilized across species boundaries. It may also be noted that, at least in some cases, enhanced growth is associated with more effective utilization of food.

Cold water temperatures are often a major problem in aquaculture in temperate climates when an unusually cold winter can severely damage both production and brood fish stocks of fish. Some marine teleosts have high levels of serum anti-freeze proteins (AFP) or glycoproteins (AFGP) which reduce the freezing temperature by preventing ice-crystal growth. Fletcher, Hew and Davies have shown that there is one class of AFGP and four classes of AFP. Most are expressed primarily in the liver and some show clear seasonal changes. Work has particularly focussed on the production of AFP from the winter flounder (*Pleuronectes americanus*), and the gene has been successfully introduced into the genome of Atlantic salmon, integrated into the germ line and passed on to F3 offspring where it was expressed in the liver. However, a number of Ala, Pro-specific endopeptidases are required for production of mature proteins and these are not present in Atlantic salmon. Furthermore, the AFP gene in winter flounder, and possibly other Arctic species, exists in many copies. Thus, much further work is required in order to develop effective antifreeze activity in Atlantic salmon. Work on AFP has also been conducted in goldfish and milkfish.

Genetic manipulation has also been undertaken in order to increase the resistance of fish to pathogens. This is currently being addressed by the use of DNA vaccines (encoding part of the pathogen genome) and antimicrobial agents such as lysozyme. An example is the injection of Atlantic salmon with a DNA sequence encoding infectious hematopoeitic necrovirus (IHNV) glycoprotein under the control of the cytomegalovirus promoter (pCMV). Challenge with the virus eight weeks later revealed that a significant degree of resistance had been achieved. The fish were still resistant and were shown to have generated antibodies three months later. Similar studies have been undertaken for other fish diseases eg. Haemorrhagic septicaemia virus (VHS) and work of this kind appears to have great potential value for fish farms. We would also draw attention to the work using a cecropin B gene from the moth *Hyaloplova cecropin.* When channel catfish transgenic for this gene were challenged with *Flavobacterium columnare* and *Edwardsiella ictaluri* survival was better for transgenics than controls

Table 2. Results in induction of GMOs in aquatic species.

Species	*Target gene*	*Int*	*Exp*	*Trans*
At. Salmon	GH	+	+	+
At. Salmon	AFP	+	+	+
Coho salmon	GH	+	+	Nd
Tilapia	tiGH	+	+	+
Tilapia	Fish GH	+	+	Nd
Carp	GH	+	+	Nd
Salmon	Glucose transporter and hexokinase	+	+	+
Tilapia	GH	+	+	Nd
At. salmon	GH	+	+	+
Carp	HGH	+	+	+
At. salmon	GH	+	+	+
Tilapia	tiGH	+	+	+
Tilapia	INT-tiGH	+	+	+
Tilapia	CSGH	+	+	+
Tilapia	INT-tiGH	+	+	+

Tilapia	ypGH	+	+	Nd
Abalone	GH	+	+	Nd
Medaka	CAT	+	+	+
Tilapia	GH	+	+	+
At. salmon	GH AFP	+	+	+
Tilapia	tiGH	+	+	+
Tilapia	Lac Z	+	+	+
Tilapia	tiGH	+	Nd	Nd
Coho salmon	GH	+	+	+
Pacific salmon	CSGH	+	+	+
Common carp	RTGH	+	+	Nd
Common carp	CSGH	+	+	+
Medaka	Lac Z	+	+	Nd
Brine shrimp	Luciferase reporter gene	+	Nd	Nd
Common carp	RTGH	+	+	Nd
Common carp	CSGH	+	+	+
Pacific salmon	CSGH	+	+	+
Rainbow trout	CSGH	+	+	+
Cutthroat trout	CSGH	+	+	+
Chinook salmon	CSGH	+	+	+
Loach	CSGH	+	+	Nd
Salmon	GH	+	+	Nd
Chinook salmon		+	Nd	Nd
Laminaria japonica	Plasmid BI221	+	+	Nd
Undaria pinnatifida	Plasmid BI221	+	+	Nd
Nigorobuna	*E. coli* beta galactosidase	+	+	Nd
Blunt snout bream	HGH	+	+	+
Common carp	HGH	+	+	+
Oreochromis niloticus	Bacterial lacZ	+	+	+
Zebrafish		+	Nd	Nd
African catfish	AFP GH	+	+	Nd
Common carp	AFP GH	+	+	Nd
Abalone	GH	+	+	Nd
Pacific salmon	GH sockeye salmon	+	+	Nd
Zebrafish	Firefly luciferase	+	+	-
Zebrafish	CSGH	+	+	+
Common carp	RTGH	+	+	+
Zebrafish	luciferase	+	-	-
Common carp	HGH	+	+	+

Tilapia	RGH	+	Nd	Nd
Zebra fish	CAT	+	Nd	Nd
Tilapia	HGH	Nd	Nd	Nd
Zebrafish	CAT	+	+	Nd
Goldfish	Neomycin resistance CAT	+	+	Nd
Northern Pike	BGH	+	+	Nd
Atlantic salmon	Winter flounder AFP	+	+	+
At. salmon	Bacterial CAT Chinook salmon GH	+	Nd	Nd
Common carp	RTGH	+	+	+
Channel catfish	RTGH	+	+	+
Northern Pike	Bacterial CAT BGH and CGH	+	Nd	Nd
Walleye	Bacterial CAT BGH and CGH	+	Nd	Nd
Zebrafish	Bacterial CAT BGH and CGH	+	Nd	Nd
Carp	Grass carp GH	+	Nd	Nd
Zebrafish		+	-	Nd
Northern pike	BGH CSGH	+	+	Nd
Channel catfish	Salmon GH	+	+	Nd
At. salmon	CSGH	+	+	Nd
Gilthead seabream	BGH and HGH	+	-	Nd
Rainbow trout	Carp alpha globin	+	+	+
Rainbow trout	BGH	+	+	Nd
Tilapia	HGH	+	Nd	Nd

Abbreviations used in Table 2 above:

At:	Atlantic salmon
GH:	Growth Hormone
AFP:	Anti-freeze Protein
Nd:	Not determined
HGH:	Human Growth Hormone
BGH:	Bovine Growth Hormone
CS:	Coho Salmon Growth Hormone
YP:	Yellowfin Porgy Growth Hormone
CAT:	Chloramphenicol Acetyl Transferase
TiGH:	Tilapia Growth Hormone
RTGH:	Rainbow Trout Growth Hormone
RGH:	Rat Growth Hormone
Int:	Integration
Exp:	Expression
Trans:	Transmission

There appears to be no published evidence for integration of vaccine DNA into the recipient genome. Nevertheless, the persistence of the DNA appears often to be relatively long which suggests some replication (not normally expected with non-chromosomal pieces of DNA). It seems desirable for the moment to regard such treated animals as "transient" GMOs rather than full GMOs.

Isolation of the Gene of Interest

Usually the gene of interest will already be available as an element of a "library" of short sections of the total genome of the donor strain or species. If this is the case the procedure followed is to multiply the gene using the PCR reaction. If, however, the gene is to be taken from a genome not previously investigated, a more complex procedure will need to be followed. The use of the technique of the Polymerase Chain Reaction (PCR) enables the gene in both the cases noted above to be multiplied to the level of several million copies needed for the generation of the construct.

Cloning the Gene of Interest

When many copies of the target gene have been generated, the gene is placed in a "construct". Once the gene of interest has been ligated enzymatically into the construct, this whole complex is ligated into bacterial plasmids, which act as "production vectors" and enable the gene to be replicated many times within the bacterial cells. The bacteria are then plated out. It is possible to tell from reporter genes whether the vector has been taken up by the bacterial cells. This usually involves some colour change in the colonies containing inserted DNA. The many times amplified DNA construct is then enzymatically cut out of the plasmids (after these have been removed from the bacterial cells) and it is ready to be used for insertion into eggs of the host species.

The Construct

A construct is a piece of DNA which functions as the vehicle or vector carrying the target gene into the recipient organism. There is a promoter region which controls the activity of the target gene, a region where the target DNA is inserted, usually some type of reporter gene to enable one to ascertain whether the target has combined successfully with the construct and a termination sequence.

The sources of these several DNA sequences may be different species although promoter and target genes would ideally be derived from the same species

As shown in Table 3, constructs have been reported from 92 studies. The number of different constructs is greater than the number of target genes used in aquaculture and a substantial research effort has been made in this area. From the early 1990s research focussed on developing "all fish" constructs in preference to using mammalian promoters.

The use of all-fish constructs has dramatic effects on expression of transgenes, e.g. Devlin *et al*., developed an all salmon gene construct which accelerates the growth of transgenic salmonids by over 11 fold. In tilapia, Maclean found that using carp beta actin instead of rat beta actin promoter led to a ten fold increase in production of hormone in transgenic animals.

Table 3. Summary of major research effort in inducing GMOs in aquatic species.

Species	*Target gene*	*Typical construct*	*Typical induction method*	*Number of studies*
Salmon spp.	GH AFP	Ocean pout AFP linearized DNA	Microinjection	17/92
Rainbow Trout	GH	Ocean pout AFP	Microinjection	14/92
Tilapia spp.	GH	Cytomegalovirus (CMV)	Microinjection	12/92
Carp	GH	Rous Sarcoma Virus Long Tandem Repeat	Microinjection	17/92
Zebrafish	Luciferase	pMTL plasmid	Microinjection	16/92
Medaka	CAT	AFP	Microinjection	11/92

Other important work suggested that the optimal stage at which the transgene is introduced might vary between cells and species eg. Garcia del Barco *et al.* using Zebrafish showed that there were differences in the regulatory requirements for cells and embryos, and suggested therefore that constructs should be assayed in both cells and embryos.

Other work shows how critical the nature of the gene construct is. Devlin *et al.* showed that using an opAFPGHc gene construct in coho salmon eggs gave rise to some alevins which had the typical brown colouration, while the remainder displayed a distinct green colouration. The results suggest that the green phenotype arose from the presence of the opAFPGHc construct and therefore could be indicative of transgene uptake/transmission. All the offspring were tested by PCR for presence of the transgene and 182 of 184 alevins were correctly assigned on this basis. However, it was found that later in development all fish turned green (the normal colour later in development) and so the transgenic fish were showing accelerated growth. Later in development it was found that most of the transgenic fish showed signs of cranial abnormality probably due to accelerated growth. While the onstruct was useful in that transgene uptake could be monitored, further work was needed to ensure that healthy fish could be produced.

Techniques for Inducing Transgenics

Transgenic fish have largely been produced through microinjection into fertilised eggs or early embryos. Electroporation of sperm has been shown to be successful in some species eg. Zebrafish Chinook salmon and Loach. Liposomes have also been utilized as vectors. Ballistic methods using microprojectiles have been investigated in *Artemia* with a view to their use in generating transgenic crustacea and also in seaweed species. "Baekonisation", an electric, flat field type of electroporation was utilized to transfer DNA into Zebrafish embryos, this method appeared to be successful but has not been taken up in the same way as other forms of electroporation and microinjection methods.

More recently the use of embryonic stem cells (ESC) as a method for inducing transgenesis has been advocated. These cells are undifferentiated and remain totipotent, so they can be manipulated in vitro and subsequently reintroduced into early embryos where they can contribute to the germ line of the host. In this way genes could be stably introduced or deleted. Despite the early success of ESC technology in mice, the uptake of the technology for fish has been slow, although early precursor cells (Mes 1) have been cultivated from Medaka and show many of the same features as mouse ESC. Studies by Hong, Winkler and Schartl and Hong, Chen and Schartl showed that 90 percent of host cell blastulae transplanted with Mes 1 cells developed into mosaic fry, and these cells became integrated into organs derived from all three germ layers, and differentiated into various types of functional cells.

Another example of new and possibly more efficient ways for gene transfer is the use of pantropic retroviral vectors. These are able to infect a wide range of host cells and have been used to infect newly fertilized Medaka eggs with a reporter gene, which appeared to become integrated into the entire germ line of some of the P1 females. In Zebrafish when retroviral infection and microinjection were compared, the two methods were equally efficient in passing the transgene into eggs, but there was wider variability in the extent of reporter gene expression among those founders that were microinjected. However, the use of retroviruses is not without problems.

The microinjection method is suitable for relatively small numbers of organisms being manipulated whereas electroporation, sperm/liposome mediation and bombardment methods are more suitable for mass treatments. The most popular method of insertion of transgenes in aquaculture is microinjection; in 92 studies reviewed from 1985 to the present, 68 used microinjection, eleven used sperm mediated methods, six used electroporation and five used both sperm mediation and electroporation. However, the problem of mosaic expression of the transgenes is common, and this gives rise to varying proportions of transgenic genotypes in the progeny.

Integration Sites

The factors determining sites of integration are still poorly understood though research in this direction is increasing. It is particularly important to gain greater accuracy in controlled site of integration because of the unpredictable effects of uncontrolled integration on resident genes. Caldovic and Hackett tested the ability of special sequences called transposable border elements from other species to confer position-independent expression of transgenes or enhance integration of transgenic constructs into fish chromosomes. Early results indicate that such elements from some species do not act as enhancers and do not improve integration frequencies. However, both avian and insect border elements were found to confer position-independent expression as judged from expression of CAT genes in F_1 fish. Hackett *et al.,* showed that co-transfer of retroviral integrase protein with transgenic DNA can accelerate and enhance the rate of integration. More studies of this type are needed to improve the success and controlled positioning of integration of transgenes in the future.

Expression of Gene

The uptake and integration of a transgene does not guarantee that the gene will express itself in the new genetic environment. Tests must be carried out to determine whether there is expression and if there is expression, at what level this takes place. Clearly, in commercial aquaculture only those transgencs expressing the target gene at a sufficiently high level will be of interest.

Inheritance of Gene

A fish which expresses the target gene at an acceptable level may not be able to transmit the gene to progeny. This is because many transgenics are mosaic individuals and unless the gonads are included in the tissues possessing the transgene the transgenic animals will not breed true. Appropriate breeding tests must, therefore, be carried out.

The high proportion of mosaic individuals is one reason why the proportions of progenies of different genotypes resulting from parents that are putatively hemizygous for a transgene do not necessarily conform to mendelian expectations. Another reason is the integration of two or more copies of the transgene at different sites in the recipient genome. Further breeding tests will be required in order to establish a pure breeding line of transgenic fish.

Field of Transgenics

Inducing transgenics is a relatively inefficient process. According to Maclean relevant variables include the species, the workers involved and presumably also the techniques. For every hundred eggs injected, a yield of about ten percent of fish testing positive for the presence of the transgene may be expected. However, only about one percent of the eggs treated will prove ultimately to be germ line positive and capable of transmission to the next generation. This figure is in line with that found for pigs, sheep, goats and cattle.

Genetic Architecture of Quantitative Characters

Most of the phenotypic characters of interest to aquaculture are quantitative rather than qualitative. It is therefore important to understand the genetic architecture of such characters. The polygenic theory of quantitative characters envisaged a fairly large number of loci each with relatively small and equal effects acting in a largely additive way on a quantitative character. The theory made no assumptions about the heritability of the character so that two discrete characters might have similar numbers of loci concerned in their determination but very different values for genetic and environmental variance when these were partitioned from the total phenotypic variance. Over the years it has indeed been observed that relatively large numbers of loci may be involved but also that effects of dominance and epistasis are frequently involved and that the magnitude of the effect produced by each locus can vary considerably.

Current informed thinking on genetic architecture is admirably described by Falconer and Mackay. It can be summarised as follows:

A typical quantitative character is likely to involve:

1. a number of loci which may reach several tens in number;
2. genes acting in ways which may be additive, dominant, epistatic and interactive with environmental factors; and
3. considerable variation in individual locus effect (including many small effects from genes whose primary effect is elsewhere on the phenotype through pleiotropic effects).

The last point is quite important as it suggests that typically a small number of loci account for a very large fraction of the variation in the character.

The work of Devlin *et al.*, which suggests that the benefits of transgenic technology in strains already subject to selection for the desired phenotype may be negligible, is relevant here. The selected strain of rainbow trout which they used might reasonably be assumed to be homozygous for favourable alleles at most of the loci of larger effect.

These observations lead inescapably to the conclusion that success in achieving the desired phenotype in transgenic animals will depend on the nature of the genetic architecture of the character concerned. Where gene action is largely additive more progress may be expected than where more complex aspects of gene action are seen.

Benefits Arising From the Use of GMOS

Aquaculture

Evidence of real benefit in terms of economically significant characters comes mainly from work on growth hormone (GH). The overall conclusion from the studies of several workers is that fish GH transgenics enjoy growth rates markedly superior to those in comparable (in some cases sibling) non transgenics. Studies have revealed enhancement of growth particularly in salmonids to an average of 3-5 times the size of non-transgenic controls with some individuals reaching as much as 10-30 times the size of controls. The economic gains to be made from use of such GMOs are obvious and transgenics must therefore be considered as a route for providing superior strains along with selective breeding. We should note that, not surprisingly, lines resulting from different transgenic events with the same construct in the same population may give different results and this has been confirmed in field trials.

The species involved include Atlantic salmon, coho salmon, Nile tilapia and interspecific hybrid tilapia. Work reported on carp and channel catfish shows less but still significant effect but, as indicated by Maclean and Laight, this may be a consequence of 1) choice of promoter sequence and 2) a background of selective breeding in the strain used. In most cases the transgenics will be hemizygous for an unknown number of copies (possibly often one) of the transgene.

There is a most interesting suggestion from the work of Martinez *et al.* using tilapia GH in *O. hornorum urolepsis* that fish hemizygous for the transgene are superior in growth rate not only to wild type sibs, but also to transgenic homozygotes. This, if a real and general effect, may be of considerable significance for the use of GH transgenics in aquaculture and the maintenance of broodstock.

Considerable interest exists in making fish transgenic for the antifreeze protein genes found in some species such as winter flounder and if the difficulties involved in securing phenotypic expression of the antifreeze phenotype in a phenotype controlled by multiple loci can be solved, the benefits would be very large.

There are also a number of other target phenotypes for which transgenics offer considerable potential. These include salinity tolerance, sterility, control of sexual phenotype, disease resistance to specific pathogens and behavioural modifications. One particularly interesting possibility is that of modifying the genome to allow greater production of omega-3 fatty acids. There are, as yet, few concrete data which can be reported but clearly there are very promising areas of work which could bring substantial benefits to aquaculture.

The introduction of a transgene is intrinsically unlikely to have only one effect on the phenotype and possible pleiotropic effects need to be considered. These could in principle, be of two kinds:

i) genuine pleiotropy manifested through, for example, dose effects in the metabolic network; and

ii) apparent pleiotropy arising from disturbance in functioning of resident genes through integration of a transgene at a specific point in the genome. Such disturbances might be favourable or unfavourable.

It will not always be easy to distinguish between genuine and apparent pleiotropy. However, Chatakondi *et al.*, and Dunham have reported favourable effects such as increased carcass yield, increased protein level, reduced fat and greater tolerance of low dissolved oxygen levels in common carp and channel catfish transgenic for rainbow trout GH. Dunham has argued, without an explicit rationale, that "disease resistance will likely be improved directly".

Possible effects of other elements in the construct such as reporter genes or antibiotic resistance genes need mention. Such cotransgenes confer no benefits and may pose significant risks (particularly with antibiotic resistance genes). Best practice would certainly require removal of such elements before commercial use of the target transgenes is started.

Other Uses of Transgenics in Aquatic Species

While the primary focus of this paper is on uses of transgenics is in improving production in aquaculture, it is worthwhile pointing out that there are several other potential uses with strong connections to aquaculture. These include living pollution monitors achieved by incorporating a pollution sensitive promoter in the transgenic animal.

A typical example would be a green fluorescent protein structural gene (GFP) driven by a metallothionin promoter. If the promoter is inactivated by heavy metal pollution the GFP is switched off and the colour change is readily visible. Another use closely related to aquaculture, is that of using fish as a production system for valuable gene products which can be extracted in a comparable fashion to similar production in mammalian species. Such products might include vitamins and work is underway to produce factor VII (one of the human blood clotting factors), in tilapia.

Use of AFP of a tangential kind includes cases where it has been used to help protect membranes from cold and freezing damage by modification of the structure of membranes *in vitro*. The ability of fish AFP's to preserve sheep and pig embryos has been demonstrated. The use of AFP's in cryopreservation of fish eggs and embryos still awaits further development. However, some initial work has been carried out by who were able to cryopreserve spermatozoa from the ornamental Japanese carp. Work exploiting AFPs generated by transgenic fish could become most useful in hatcheries in future in order to preserve transgenic lines and to supply new hatcheries and farms with suitable stocks.

Commercial Significance

The demand for fish is increasing year on year and the yield from capture fisheries is declining. Thus, although aquaculture production is increasing the market for further expansion in aquacultural production is likely to be very good for many years to come.

An OECD view was that the time scale from 1995 for GMOs in salmon to be commercialized would be 15 years and that for tilapia would be five years. As matters stand at present the estimates for both species would lie between the two figures given. It is reported that Atlantic salmon transgenic for a Chinook salmon GH gene are being considered for approval in aquaculture in the USA.

The data available on GH transgenics suggest that the monetary benefits to be obtained from use of these fish will be large. For comparison, the use of the single step genetic change represented by monosex genetically male tilapia (GMT) in Nile tilapia (though this is not a GMO) increased production by almost 30 percent and effectively doubled the net income, from this source, of Philippine farmers growing it. Nevertheless it is sensible to recognize that the benefits of use of GMOs are not always clear cut, at least in crop plants in the USA.

Risk Factors of GMOS

There are a number of publications which address this issue. Maclean and Laight and Dunham have produced very useful reviews which discuss many of the points raised in this paper.

In our view the most important areas of risks which need to be considered in the use of transgenics are:

1. human health
2. biodiversity
3. animal welfare

4. poor communities

In each of these categories there exists a multiplicity of pathways by which effects could, in principle, be brought about. Rational and responsible assessment of risk requires that the following properties are all considered:

1. source of the DNA of the target gene;
2. source of the non target DNA segments of the construct used;
3. site(s) of incorporation of the transgene within the recipient genome;
4. product of the transgene;
5. interaction of the transgenic product with other molecules in host and consumer;
6. possible molecular changes in transgene product during processing;
7. pleiotropic effects of transgene;
8. tissue specificity of transgenic expression; and
9. numbers of transgenic organisms capable of interacting with natural systems).

Human Health

The risks to health will depend upon all of the factors listed above. In practical terms the most important of these are likely to be the source of the DNA and the nature of the product.

The great majority (98 percent) of dietary DNA is degraded by digestive enzymes relatively quickly but use of viruses (disarmed or otherwise) as vectors, must increase the risk factor significantly as these are organisms which are adapted to integrating into host genomes and some represent risk factors for cancer induction. The work of Zhixong Li *et al.* who induced leukaemia by using retroviral vectors in making transgenics for a commonly used marker gene in mice and a recent report of leukaemia induction in a child undergoing gene therapy for x-SCID using a retrovirus show that this is not a trivial risk. Arguments about risks and benefits attached to this form of gene therapy are current.

At the other extreme the use of autotransgenics must be seen as posing a risk which is orders of magnitude lower than that for allotransgenics and probably negligible. The major risk from the production of the transgene will lie in the use of novel proteins or other molecules produced by the transgenic organisms. Either in the native form or, following modifications in the human body, such molecules could be inimical to human health (e.g. through allergies). It would seem sensible to avoid the use of such substances except where strictly necessary and under rigorous control.

Other potential risks may lie in incorporation of transgenic DNA into the genomes of resident gut microflora (though this is likely to be very improbable) or a change in the pathogen spectrum of the transgenic fish leading to it hosting a new pathogen which happens to be also a human pathogen.

Biodiversity

The extent of aquatic diversity is both extremely large and relatively poorly understood. This means that the task of estimating the risks to aquatic biodiversity at all of its levels from the use of GMOs or indeed, any genetically distinctive strain used in aquaculture is monumentally large. Aquaculture has a further problem in that the (almost always unintended) escapes of genetically distinct farmed fish are unpredictable and often large in numbers. Stenquist in discussing transgenics in open ocean aquaculture, quotes some relevant figures. Thus, 15 percent escapes for Atlantic salmon, escapes of 150 000 salmon and 50 000 trout in Chile and catch statistics for Atlantic salmon off Norway in which 15-20 percent of the fish caught were of farmed origin. In Scotland an escape of 100 000 Atlantic salmon was reported recently. It is clear that escapes of these magnitudes pose considerable problems and it is not surprising that in some parts of Norway fish of farmed origin represent a majority of the animals fished

The major focus of attention in the literature lies, understandably, upon the effects of escapes upon natural populations of the same species, but we must always bear in mind possible impacts across an assemblage or ecosystem as a whole. The first general point to make is that there is, in principle, no difference between the biodiversity risks from escapes of GMOs and from fish genetically improved in some other way, e.g. by selective breeding or (in some respects) from exotic species.

The second general principle is that such genetically improved forms including GMOs, are developed for a specific set of environmental circumstances in which they enjoy an advantage conferred by human decisions. In nature, however, such genetically distinct forms may legitimately be regarded as mutant forms of the wild type. A considerable body of genetical knowledge tells us that the probability of survival of mutant forms is extremely low because they are disadvantaged in viability and/or fertility under natural conditions. Thus, for example, in the genetically distinct farmed Atlantic salmon in Norway the males are very much less successful than wild males in securing mates.

However, it must be conceded that in species like salmon where the farmed populations outnumber the wild populations by orders of magnitude, the effects of escapes of any genetically distinct genotype upon natural populations may be both deleterious and of significant size simply as a result of "swamping"

An interesting model of the effects on a medaka (*Oryzias latipes*) population of transgenic release has been produced by Muir and Howard using estimates of juvenile and adult viability, age at sexual maturity, female fecundity, male fertility and mating advantage. They were able to demonstrate that the transgene would spread in natural populations, despite low juvenile viability, if transgenes have sufficient high positive effects on other fitness components. It has been argued that this might lead to extinction but the selective pressure for recombinant genomes with higher viability would be expected to be immense.

Maclean and Laight simulated the changes in frequency of a transgene expected with different scenarios embracing a range of selective values including heterozyote advantage. They note that "repeated small introductions [of the transgene] can have an effect on... frequency... since the frequency of advantageous alleles rises much more rapidly than if a single large introduction is considered".

A major problem in assessing risk to natural populations is that of scale. Even if farmed fish are at a selective disadvantage in natural conditions, the ratio of wild:farmed numbers may in some areas, be relatively small. In these situations significant modification of the "native" population and its role in the ecosystem is inevitable.

Whilst not providing a completely satisfactory answer, there is little doubt that making farmed fish sterile would go a long way towards reducing the pressure upon such threatened ecosystems. A number of research efforts to develop systems for sterile fish production are being made. The techniques include triploidisation, antisense transgenics, ribozymes and gene targeting.

Provided that the best containment measures (physical and biological) are adopted, in our opinion, in general risks to biodiversity by GMOs *per se* are probably extremely small, but in specific cases, the risks and consequences may be large. As a general rule and adopting a precautionary approvah, it is, however, clear that each individual case needs careful study and appraisal and the best possible containment measures before approval for uptake into commercial production is given.

Animal Welfare

The direct or indirect effects of transgenesis upon the welfare of fish GMOs in aquaculture are very poorly understood. In part, no doubt, this is because notions of cruel or unnatural treatment in mammalian species translate, for a variety of reasons, imperfectly to fish. Nevertheless, as life forms with highly developed nervous systems and with a range of behavioural phenotypes which flow from this, fish qualify for welfare consideration.

There are a few studies which bear on this. Thus, for example, Devlin *et al.* reported changes in colouration, cranial deformities and opercular overgrowth and lower jaw deformation in coho salmon transgenic for AFP and GH. After one year of development anatomical changes due to growth of cartilage in the cranial and opercular regions were more severe and reduced viability was evident.

The larger body of data on species farmed terrestrially shows dysfunctional development leading to acromegaly, lameness and infertility in some GH transgenics in pigs and sheep. However, in pigs dietary modification influencing nutritional levels of zinc proved successful in avoiding such abnormalities.

We have been unable to find systematic data on the incidence, in fish GMOs, of effects such as those described by Devlin *et al.* and this is probably because animal welfare is not sufficiently widely recognised as an issue in relation to the use of GMOs. This is well illustrated in the otherwise comprehensive and balanced review by Sin in which the section on ethical issues contains no reference to animal welfare. Nevertheless, if GMOs are to be used in aquaculture (and there are weighty arguments for so doing), concerns on this issue will need to be properly satisfied. The Royal Society report devotes a significant amount of space to this issue.

Poor Communities

This term rather than poor countries is used because all poor countries contain rich people and rich communities. The possible economic disadvantages of use of transgenics centre on two issues:

Dependence on external agencies for seed fish

If transgenic fish become widely grown because they are much more efficient, and if special broodstock are required to produce fry for on-growing to adults, which, cannot be used as broodstock, a dependency is created. This dependency may be benign or oppressive, depending on the arrangements made for seed supply.

Intellectual property rights

This is a very difficult issue indeed. Since genes may now be patented and therefore, enjoy commercial value, the opportunities for dispute about equitable treatment of stakeholders in cases where ownership of genes and strains is contested, are legion.

A recently published report states that developing countries are frequently disadvantaged in the use of, and access to, IPR because of increasingly protective attitudes taken by owners of IPR. However, the report also indicates that developing countries are very heterogeneous in respect of their ability to use and develop IPR.

Regulation Policies

The extent to which GMOs are perceived as desirable for aquaculture and food supply has been probed by Bartley and Hallerman on a questionnaire basis. The responses were generally positive in terms of exploring the potential of transgenesis. It is clear however, that in human populations as a whole there is a severe deficiency of knowledge appropriate for making informed decisions about the value of GMOs.

The task of instituting and managing well thought out, responsible and scientifically sound measures is made more difficult by the frequently irresponsible and inaccurate media treatment of GMOs on the one hand and the cavalier pronouncements by some authorities on the other hand, e.g. that no distinction need by made in labelling between food derived from GMOs and from non-GMOs. Such extreme differences in attitude tend to inflame public opinion.

The need for incorporation of risk assessment and risk management procedures relating to use of GMOs in aquatic species has been well brought out by Hallerman and Kapuscinski.

According to Hallerman and Kapuscinski, "as a generality among developed countries at least, the public will support biotechnology if it yields a healthful product in an environmentally sound manner. This statement encapsulates implicitly a most significant factor - that in considering the benefits and disadvantages of GMOs, perception by the general public is a most important factor in shaping attitude of regulatory agencies".

There is at least a suspicion that some contemporary science may also be affected by perceptions. A recent furore over the alleged presence, in traditional varieties of maize in Mexico, of transgenes derived from modern strains of cultivated maize points to this. The matter is still not completely resolved but has led to *Nature* declaring it should not have published the original paper because of technical inadequacies.

It is perhaps dangerous to extrapolate directly from agronomy to aquaculture but it is worth noting that, in the USA, which accounts for more than two thirds of the acreage planted to crop GMOs, moves to institute a much tighter regulatory framework are in train. Five bills were introduced in May 2002 in the House of Representatives in Washington to cover legal protection for farmers, increase GM food safety, require labelling for foods containing, or produced with, GMOs, address developing country issues and assign liability for damages. These measures appear to have sprung, at least in part, from serious dissatisfaction on the part of some farmers with GM crop plants. It is, however, relevant to note that the dominant position enjoyed by several large multinational agrochemical companies in providing both seed and agrochemicals is not reflected in the aquaculture industry which is far more fragmented. Nevertheless, the indications are already present that there is a move towards fewer and larger companies within the sector and this could have consequences for the control and use of GMOs in aquaculture species.

A fair number of countries have instituted regulatory arrangements for the culture, release and dietary utilisation of GMOs, but with considerable differences in the approaches and restrictions employed.

At the international level the Cartagena Biosafety Protocol of the Convention on Biological Diversity provides a comprehensive and rigorous framework for regulation of protection of biodiversity and argues that this be done "taking into account risks to human health and specifically focussing on transboundary movements". The great majority of nations are signatories to the convention. What is less clear is the extent to which, at national and regional levels, the protocols set out are implemented and implementable through domestic legislation and effective sanctions.

Among its provisions, the Cartagena protocol has a detailed treatment of risk assessment in relation to the protection of biological diversity. Among the important and relevant general principles are:

1. "Risk assessment should be carried out on a case by case basis."
2. "Risks associated with LMOs should be considered in the context of the risks posed by the non-modified organisms in the receiving environment."

and in the [comparable] methodology of risk assessment:

1. "An estimation of the overall risk posed by the LMO based on an evaluation of the likelihood and consequences of the adverse risks being realized."
2. "A recommendation as to whether or not the risks are acceptable or manageable, including, where necessary, identification of strategies to manage these risks."

We are, thus, faced with a complex array of stakeholders; breeders and growers, geneticists, multinational companies, supermarkets, consumers, politicians and the media. All have their own agenda to follow, but critically important to all of them and to a reasonable resolution of the current conflict is the availability and use of full, sound and accurate information, particularly in the context of framing of appropriate regulations affecting GMOs.

References

Beardmore, J.A.B. 1997. Transgenics: autotransgenics and allotransgenics. *Transgenic Research* 6: 107-108.

Beardmore, J.A.B. 1998. Genetic impacts in aquatic species. *Biol. Jb. Dodonaea* 65: 69-80.

Fjalestad, K.T., Gjedrem, T. & Gjerde, B. 1993. Genetic improvement of disease resistance in fish: An overview. *Genetics in Aquaculture* IV. Eds G. A. E. Gall and H. Chen pp.65-74.

Gjoen, H.M. & Bentsen, H.B. 1997. Past, present and future of genetic improvement in salmon aquaculture. *ICES Journ. Mar. Sci.* 54: 1009-1014.

Zhu, Z. 1992. Generation of fast growing transgenic fish: Methods and mechanisms. In *Transgenic Fish.* Eds. C. L. Hew and G. L. Fletcher. Singapore, Singapore World Scientific, pp.92-119.

4

Fish Genetics and Breeding

Fishes are paraphyletic group of organisms that consist of all gill-bearing aquatic craniate animals that lack limbs with digits. Most fish are ectothermic allowing their body temperatures to vary as ambient temperatures change, though some of the large active swimmers like white shark and tuna can hold a higher core temperature. Fish are an important resource for humans worldwide, especially as food. Commercial and subsistence fishers hunt fish in wild fisheries or farm them in ponds or in cages in the ocean. They are also caught by recreational fishers, kept as pets, raised by fishkeepers, and exhibited in public aquaria.

Fish flesh is an important source of high grade animal protein. With a growing world population and rising expectations for living standards, the pressures on fish both as a source of food and for its byproducts can only increase in the future. At present fish and aquatic animals constitute 17 percent of the total animal protein in the human diet. This statistic conceals wide regional differences in fish consumption. For example, 32 countries obtain 34 percent or more of their animal protein from seafood. On the African continent, ten countries obtain over 40 percent of their protein from fish. In 21 countries on the same continent over half of the fish landed comes from inland lakes and rivers, and in 13, the fish supply is entirely from inland sources.

The potential for further growth of natural marine and freshwater fisheries is limited. Already the total world harvest is 15–20 million tons lower than it might have been had fishing been strictly controlled through scientific management, and 25 valuable fisheries have been seriously depleted by overfishing. Other factors which limit the contribution of marine fish to the diet include problems of distribution, storage and cultural inhibitions to its consumption.

Intensive fish farming can, in principle, overcome two of these limitations. Fish farming activities may often be sited near to the markets in which the product is sold. They also avoid the depletion of biological resources except where still dependent on the collection of seed from the wild. Fish farming not only allows the products to be delivered directly to their markets but it also represents both a renewable and inexhaustable resource. Aquaculture is still in a relatively early stage of development and the production of new strains and races of food fishes will require careful management of available genetic resources. The lesson learnt from extensive breeding in crop plants and in domestic animals is that a narrowing in the genetic base of the species is inevitable. In such processes, genetic determinants which are likely to be lost at an early stage include those controlling disease-resistance and fitness in marginal environments. It is hence prudent for breeders to be especially mindful of the need to protect and preserve, at the earliest stage possible, broad genetic diversity within those species most likely to come under intensive breeding pressure.

Fisheries not only provide a significant portion of the protein available for human consumption -they are also an economically significant activity, providing jobs and investment opportunities, and, for many countries, a means of improving the balance of international trade.

Certain species of fish are, by virtue of special or unusual biological features, useful as experimental animals and as a source of biochemical and pharmacological agents. Often species and genetic traits to be preserved are not the same as those for food fishes. Since only a handful of the 25000 plus species of fish have been subject to scientific scrutiny the extinction of any species may represent a potential loss of an economic resource.

What is Genetics ?

Genetics is the science of genes, heredity, and the variation of organisms. Heredity and variations form the basis of genetics. In modern research, genetics provides important tools for the investigation of the function of a particular gene, e.g., analysis of genetic interactions. Within organisms, genetic information generally is carried in chromosomes, where it is represented in the chemical structure of particular DNA (deoxyribonucleic acid) molecules.

Genetics deals with the molecular structure and function of genes, gene behavior in context of a cell or organism (e.g. dominance and epigenetics), patterns of inheritance from parent to offspring, and gene distribution, variation and change in populations, such as through Genome-Wide Association Studies. Given that genes are universal to living organisms, genetics can be applied to the study of all living systems, from viruses and bacteria, through plants and domestic animals, to humans (as in medical genetics).

The fact that living things inherit traits from their parents has been used since prehistoric times to improve crop plants and animals through selective breeding. However, the modern science of genetics, which attempts to understand the process of inheritance, only began with the work of Gregor Mendel in the mid-19th century. Although he did not know the physical basis for heredity, Mendel observed that organisms inherit traits by way of discrete units of inheritance, which are now called genes.

Genes correspond to regions within DNA, a molecule composed of a chain of four different types of nucleotides—the sequence of these nucleotides is the genetic information organisms inherit. DNA naturally occurs in a double stranded form, with nucleotides on each strand complementary to each other. Each strand can act as a template for creating a new partner strand. This is the physical method for making copies of genes that can be inherited.

The sequence of nucleotides in a gene is translated by cells to produce a chain of amino acids, creating proteins—the order of amino acids in a protein corresponds to the order of nucleotides in the gene. This relationship between nucleotide sequence and amino acid sequence is known as the genetic code. The amino acids in a protein determine how it folds into a three-dimensional shape; this structure is, in turn, responsible for the protein's function. Proteins carry out almost all the functions needed for cells to live. A change to the DNA in a gene can change a protein's amino acids, changing its shape and function: this can have a dramatic effect in the cell and on the organism as a whole.

Although genetics plays a large role in the appearance and behavior of organisms, it is the combination of genetics with what an organism experiences that determines the ultimate outcome. For example, while genes play a role in determining an organism's size, the nutrition and health it experiences after inception also have a large effect.

Genes and Chromosomes

Genes are located on structures called "chromosomes," which are located in the nucleus of every cell. Although there are some exceptions, chromosomes typically occur as pairs in animals, and for practical fish culture management, fish can be considered to be diploids. Some rare species of fish are triploids, but none are aquacultured species. Some species of fish are tetraploids, notably the salmonids and catostomids; however, for practical breeding work these species behave as if they were diploids, so they can be considered to be diploids.

Because chromosomes occur in pairs, each gene also occurs as a pair. There are some exceptions in fish with morphologically distinct sex chromosomes, but few fish have such chromosomes; consequently, for practical fish breeding work, genes can be considered to occur in pairs. One chromosome of each pair comes from a fish's mother, while the other comes from its father; this means one gene from each pair comes from the mother, while the other comes from the father. The two chromosomes that form a pair are called "homologues."

Meosis

Meiosis is the process during which the primary gametocytes develop into eggs or sperm. A number of important cellular changes occur during meiosis, but for a geneticist, the important ones are those that affect the genes and chromosomes. Meiosis consists of two cell divisions.

The first meiotic division is called the "reduction division," and the second is called the "equational division." Prior to the reduction division, the homologues that form each chromosome pair replicate. The replicated homologues do not separate; they are joined together at the centromere, and the joined replicated homologues are called "sister chromatids" (Figure 1).

The replication of the chromosomes is usually uneventful in that they replicate themselves perfectly. However, mistakes in the replication process occur. These mistakes are called "mutations." The mutation rate for each gene is low, ranging from one in 10,000 replications to one in 100,000 replications. When a mutation occurs, the erroneous copy of the gene is often capable of producing an alternate version of the original phenotype.

Most mutations are harmful, but some mutations create new versions of genes, and these new variants produce new or improved phenotypes that increase survival; this can lead to the evolution of new species. If replication were perfect, life would not have evolved beyond the single-celled state. Mutations also help explain some of the consequences of inbreeding.

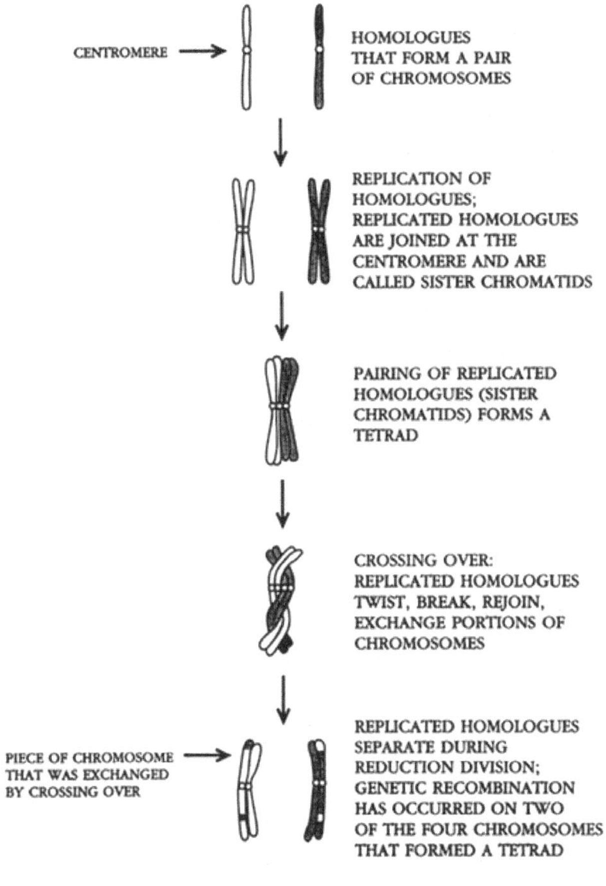

Figure 1. Schematic diagram of crossing over, which occurs in the initial phase of the reduction division during meiosis.

Prior to the reduction division, the homologues that form each pair of chromosomes are replicated. This is when mutations occur. In this figure, the homologue that came from the mother is white, while the homologue that came from the father is gray. The replicated homologues are joined at the centromere and are called sister chromatids. During the initial phase of the reduction division, the replicated homologues pair and form a bundle of four chromosomes called a tetrad. The chromosomes that form a tetrad typically twist around each other, break, and rejoin. Pieces of chromosome can be exchanged between maternal and paternal homologues when this occurs. This process is called crossing over.

During the early phase of the reduction division, the replicated homologues that form each chromosome pair come together. Because each homologue is composed of a pair of sister chromatids, each bundle is composed of four chromosomes called a "tetrad." The chromosomes that form a tetrad typically twist around each other and break. When the chromosomes rejoin, pieces can be exchanged between homologues. When this happens, pieces of chromosomes, along with the genes that are located along those pieces, are transferred from one homologue to the other. This process is called "crossing over" (Figure 1).

Crossing over is one of the most important biological processes, because it greatly increases genetic variability. In inbreeding, crossing over is important because it can affect the amount of inbreeding produced during one form of chromosomal manipulation. Crossing over tends to occur at particular locations along each chromosome. Each gene has a given cross over frequency. Genes located near the centromere cross over less frequently than those further away. Gene that are located near each other on a chromosome cross over as a unit and are called "linked"; the closer they are, the more tightly linked they are. Consequently, crossing over causes actual levels of inbreeding to deviate from predicted levels during meiotic gynogenesis.

During reduction division, the chromosome number is reduced from the diploid state to the haploid state. The homologues that form each chromosome pair separate and are parceled into the two daughter cells. When the chromosomes divide, they are not parceled into maternal and paternal sets. The direction in which each replicated homologue of each chromosome pair goes is random and independent of that for all other chromosomes. This process is called "independent assortment." Although each secondary gametocyte contains a haploid set of chromosomes, each chromosome is duplicated.

The second meiotic division is called the "equational division." During the equational division, the replicated homologues in each secondary gametocyte separate and are parcelled into either two sperm cells or into the egg and second polar body. This produces haploid gametes; each gamete contains a single chromosome from each pair, which means that it contains a single copy of each gene.

Phenotype and Genotype

Each gene contains the chemical blueprint for the production of a protein. This protein either forms or helps produce a specific phenotype, such as body colour or weight. A phenotype is the physical or chemical expression of what a gene produces; it is either observed and described or it is measured. A gene can occur in one or more forms. Alternate forms of a gene are called "alleles." In a population, a gene can have one to perhaps a dozen alleles. If there is only one allele in the population, the gene is said to be "monomorphic." If there are two or more alleles at a locus, the gene is said to be "polymorphic."

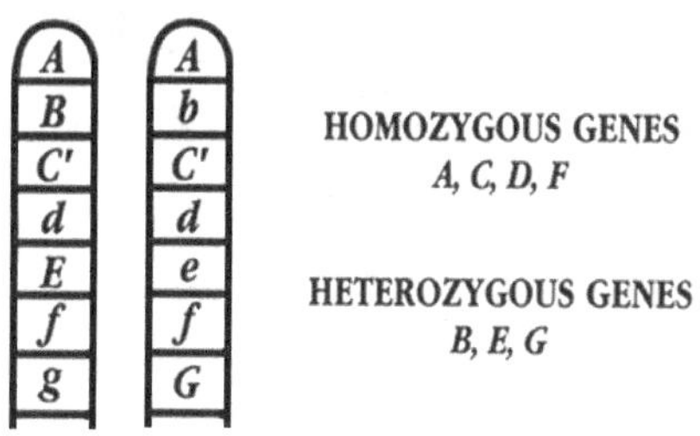

Figure 2. A small section of a pair of chromosomes, showing seven genes. Four of the genes exist in the homozygous state (A, C, D, and F), while three exist in the heterozygous state (B, E, and G)

Monomorphic genes are not very interesting, because there is no genetic or phenotypic variance associated with that locus. On the other hand, polymorphic genes are of great interest to a geneticist. When more than one allele exists at a locus, there is genetic variance, which produces phenotypic variance; and this can be exploited by breeding programmes.

The genetic make-up of each fish is called its "genotype." Because chromosomes occur as pairs, each gene occurs as a pair; this means the genotype is a paired entity (Figure 2). If the copy of the gene that a fish inherits from its father is the same as the copy that it inherits from its mother, that gene is an identical pair of alleles, and the fish is said to be "homozygous" at that locus (Figure 3). If the copy of the gene that a fish inherits from its father is different from the copy that it inherits from its mother, the gene is a non-identical pair of alleles, and that fish is said to be "heterozygous" at that locus. Consequently, every gene can exist in one of two genotypic states: homozygous and heterozygous.

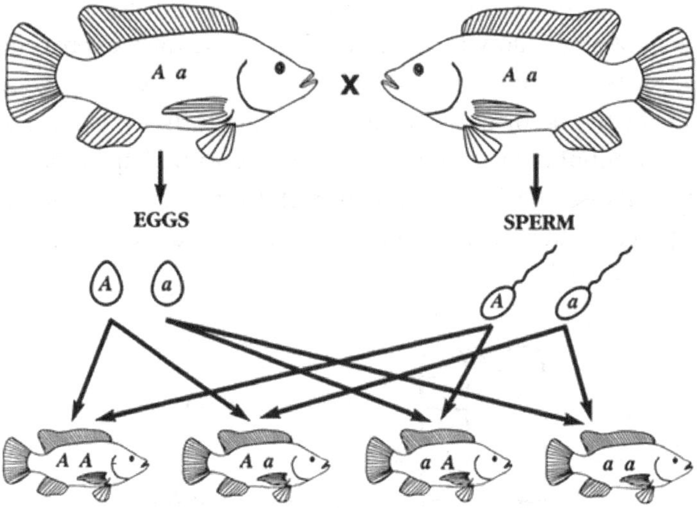

Figure 3. How homozygous and heterozygous fish are produced. Progeny with AA and aa Genotypes are Homozygous; they inherited identical alleles from both parents; Progeny with A a and a A Genotypes are heterozygous; they inherited a different allele from each parent

In this example, both parents were heterozygotes *(Aa)*, so each produced two types of gametes: half the eggs and half the sperm carried an A allele, while the other half carried an a allele. Some progeny were homozygous *(AA* or *aa)* because they inherited the same allele from both parents. Some progeny were heterozygotes *(Aa)* because they inherited a different allele from each parent.

The reason why it is important to make a distinction between the homozygous and heterozygous states, is they often produce different phenotypes. The number of phenotypes produced by a gene depends on the mode of gene action and the number of alleles.

A fish's genotype produces its phenotype. Consequently, if a breeder wants to conduct a breeding programme to control or alter phenotypic frequencies (using inbreeding or preventing it are both breeding programmes), he should understand how the genotype produces the phenotype.

Qualitative Phenotypes

Breeders divide phenotypes into qualitative (descriptive) phenotypes and quantitative (measured) phenotypes. Qualitative phenotypes are those that are described, such as body colour. Because these phenotypes are descriptive, individuals segregate themselves into descriptive categories: pink vs. normally pigmented; gold vs. bronze vs. black; normal vs. deformed.

In general, qualitative phenotypes are controlled by one or two genes. The alternate forms of a phenotype (pink vs. normally pigmented or normal vs. deformed) are produced by the alternate forms of a gene (the alleles). Most qualitative phenotypes that have been described genetically in fish are controlled by a single autosomal gene with two alleles (autosomes are the pairs of chromosomes that males and females share in common; the pairs of chromosomes that are not the pair of sex chromosomes). Because of that, this discussion will concentrate on these phenotypes, especially since this type of inheritance helps explain how inbreeding works.

In general, genes express themselves in either an additive or a non-additive manner. When genes express themselves in a non-additive manner, one allele is dominant, while the other is recessive. The terms "dominant" and "recessive" carry no value judgement; the terms simply refer to the way the alleles produce their respective phenotypes.

There are two types of dominant alleles. When the dominant allele exhibits complete dominance, it always produces its phenotype, while the recessive allele can produce its phenotype only in the absence of the dominant allele. Thus, a gene that produces two phenotypes via complete dominance has three genotypes, but the genotypes produce only two phenotypes:

Genotype	*Phenotype*
Homozygous dominant	Dominant
Heterozygous	Dominant
Homozygous recessive	Recessive

This type of gene action plays an important role in the genetics of inbreeding, and it helps explain inbreeding depression.

A second type of dominance occurs when the dominant allele is incompletely dominant; i.e., the dominant allele always produces its phenotype, but it is unable to completely suppress the recessive allele in the heterozygous state. When this happens, a gene with incomplete dominant gene action has three genotypes, and each genotype produces a unique phenotype:

Genotype	*Phenotype*
Homozygous dominant	Dominant
Heterozygous	Near-dominant
Homozygous recessive	Recessive

The heterozygous genotype produces a phenotype that resembles but is slightly different from the dominant phenotype.

When the mode of gene action is additive, neither allele is dominant, and each allele always produces its phenotype in a unidirectional step-wise manner. This means the heterozygous phenotype is intermediate between the two homozygous phenotypes. When this occurs, a gene with additive gene action produces a unique phenotype for each genotype:

Genotype	*Phenotype*
Homozygous allele 1	Homozygous phenotype 1
Heterozygous	Heterozygous
Homozygous allele 2	Homozygous phenotype 2

Only one additive gene has been discovered in fish; it controls golden, palomino, and normal body colours in rainbow trout, *Oncorhynchus mykiss*. All other qualitative phenotypes that are produced by a single autosomal gene are controlled either by complete dominance or incomplete dominance.

Quantitative Phenotypes

Quantitative phenotypes are those that are measured, such as length, weight, and fecundity. The important production phenotypes are quantitative ones, although some qualitative phenotypes are quite important, and can greatly increase the value of a crop. Because quantitative phenotypes are measured, each phenotype is a single category, such as length. Individuals do not get segregated into alternate phenotypic categories, such as long vs. short; instead, they are distributed along a continuum, and differences among individuals are determined by the unit of measure that is used to assess the phenotype: millimeters, grams, etc.

Because each fish's phenotype is determined by a measurement, quantitative phenotypes form what are called "continuous distributions," which are described by the population's mean and the distribution about the mean. In a population, these phenotypes form what are called "normal" or "bell-shaped" distributions.

The reason why quantitative phenotypes exhibit continuous distributions and why individuals do not get segregated into descriptive categories is that quantitative phenotypes are far more complicated genetically than qualitative phenotypes. Each quantitative phenotype is controlled by dozens to hundreds of genes, each of which makes a small contribution to the production of the phenotype. The exact number of genes is usually never known. Additionally, each phenotype is strongly influenced by the environment, and the influence varies both from family to family and from individual to individual. The simultaneous action of many genes and the environmental effects creates single phenotypic categories (e.g., length), in which the only way to describe an individual is to measure it.

Because each quantitative phenotype is controlled by numerous genes, as well as environmental variables, the only way to work with these phenotypes is to analyze the phenotypic variance that exists in the population and to divide the phenotypic variance into its component parts. Phenotypic variance (V_P) is the sum of the genetic variance (V_G), environmental variance (V_E), and genetic-environmental interaction variance ($V_{G\text{-}E}$) components:

$$V_P = V_G + V_E + V_{G\text{-}E}$$

When conducting a breeding programme, a geneticist tries to exploit V_G. Three distinct types of genetic variance combine to make V_G, and it is important to know what they are, because different breeding programmes are needed to exploit each type. Genetic variance is the sum of additive genetic variance (V_A), dominance genetic variance (V_D), and epistatic genetic variance (V_I):

$$V_G = V_A + V_D + V_I$$

These components of genetic variance do not refer to additive, dominance, and epistatic gene action; they refer to specific components of phenotypic variance that are produced by the entire genome, not that produced by one or two genes.

The major components are V_A and V_D. Epistatic genetic variance is usually considered to be unimportant, because it is difficult to exploit and improvements that occur by exploiting V_I plateau quickly.

Additive genetic variance is the component that is due to the additive effect of all the fish's alleles taken independently; i.e., the sum of the effects that each allele makes to the production of the phenotype. Additive genetic variance is the most important component of V_P, and the percentage of V_P that is controlled by V_A is called "heritability" (h^2):

$$h^2 = V_A/V_P$$

Additive genetic variance is the genetic component that can be exploited by selective breeding programmes. In general, phenotypes with h^2's > 0.25 can be improved efficiently by individual (mass) selection; when h^2's are $<$ 0.15, family selection is needed to improve a population. When h^2 is < 0.10, family selection is often ineffective, although some gains can be made when inbreeding is combined with family selection.

Dominance genetic variance is the other major component of V_G. Dominance genetic variance is the component that is due to the sum of each interaction that exists between the two alleles at each locus. Because V_D is produced by the interaction of the alleles at a locus, V_D cannot be inherited from either parent. Dominance genetic variance is a function of the diploid state, and offspring inherit alleles that exist in the haploid state from each parent. Dominance genetic variance is created at fertilisation when the haploid set of chromosomes from the mother pairs with the haploid set of chromosomes from the father. When this occurs, each gene exits in the paired state, and V_D is created. Consequently, V_D is destroyed by meiosis, and it is recreated anew and indifferent combinations at fertilisation.

The breeding programme needed to exploit V_D is crossbreeding. When h^2 is < 0.15, crossbreeding is often prescribed to exploit V_G.

Inbreeding

Inbreeding is the mating of relatives. Genetically, all inbreeding does is increase homozygosity in the offspring. The increase in homozygosity occurs because related fish share alleles though one or more common ancestors; i.e., the parents may carry a copy of an allele that both inherited from a common ancestor. When relatives mate, the alleles that they share because of their common ancestor(s) can be paired in their offspring. This produces offspring that are more likely to be homozygous at one or more loci.

The mating of unrelated fish also produces offspring that have homozygous genes. Additionally, an inbred fish looks the same as one with no inbreeding; there is no distinguishing mark that separates fish into inbred vs. non-inbred categories. There are two types of homozygosity are identical; there is no genetic difference. The only difference between inbred and non-inbred fish is that the homozygosity of each is created in different ways, but this distinction is most important.

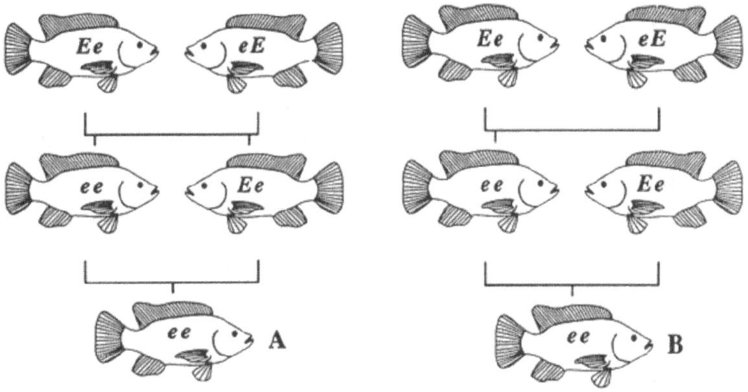

Fish A is homozygous by descent; *Fish B is homozygous in kind;*
Fish A is inbredfish B is not inbred

Figure 4. Pedigrees of two fish with identical genotypes (ee), one of which is inbred and one of which is not

Inbred fish are homozygous because they have genes where the two alleles are identical by descent; non-inbred fish are homozygous because they have genes where the two alleles are identical because they are alike in kind (Figure 4). Inbred fish are homozygous at a locus because they inherited identical copies of an allele from both parents that the parents, in turn, inherited from a common ancestor. Non-inbred fish are homozygous at a locus because they just happened to inherit an identical pair of alleles from their parents. This difference in the type of homozygosity is not physical or chemical. The only difference is the paths that the alleles took before they were paired to create the homozygosity; i.e., how the alleles were inherited.

Since there is no difference, why do geneticists make a distinction between the two types of homozygosity? Why is inbreeding of any concern? The reason is that related fish are more alike genetically than unrelated fish. This means that when relatives mate, they produce offspring that have more homozygous loci than is the case when unrelated fish mate.

The mating of relatives produces offspring that tend to be more homozygous than the population average. This increase in homozygosity is called "inbreeding," and the coefficient of inbreeding (F) is a measure of how much more homozygous a fish is than the population average. The coefficient of inbreeding does not measure how many homozygous loci the fish has; it simply quantifies the percent increase in homozygosity. On a gene-by-gene basis, F is the probability that the two alleles will be identical by descent. Since F measures the percent increase in homozygosity, the same level of inbreeding can produce different amounts of homozygosity in different populations, depending on the level of homozygosity that existed before inbreeding began.

The coefficient of inbreeding merely quantifies the increase in homozygosity that has occurred over a specific time interval due to identity by descent. Additionally, F is based on probability, so a given value of F is an average value; consequently, a given value of F means more homozygosity will have been produced in some fish, while less will have been produced in others.

The amount of inbreeding that is produced when relatives mate depends on the closeness of the relationship between the parents. Close relatives share many alleles inherited from common ancestors, so when they mate, they produce offspring with high levels of inbreeding. Distantly related individuals (e.g., fourth cousins; fifth cousins) share comparatively few alleles from a common ancestor, so when they mate, they produce offspring with comparatively little inbreeding.

Because inbreeding increases homozygosity, it changes genotypic frequencies, increasing the percentage of homozygous genotypes, while shrinking the percentage of heterozygous genotypes. If regular systems of inbreeding are conducted, the breeding programme creates a series of inbred families which do not interbreed. This subdivides a population into numerous lines, which further increases genotypic variance.

While inbreeding changes genotypic frequency, it does not change gene frequency. Selection, genetic drift, migration, and mutation are the evolutionary and breeding forces that change gene frequency. Inbreeding itself does not alter gene frequency, but by altering genotypic frequency inbreeding can, theoretically, accelerate selection. During an inbreeding programme, genotypic frequencies can be drastically altered if relatively few inbred families are maintained. When this occurs, gene frequencies will also change.

Uses of Inbreeding

Inbreeding is a breeding programme that can be used to produce superior animal and plant brood stock, and it can also be used to produce genetically improved animals and plants for grow-out. Inbreeding is the breeding programme that is often used to create new breeds or varieties that breed true for "type"; i.e., a particular body conformation or set of qualitative phenotypes. Linebreeding is a form of inbreeding that is used to increase an outstanding animal's contribution to a population. The use of herd bulls is a form of inbreeding that is used to quickly improve a population. Inbreeding can be used as a type of progeny testing to create defect-free animals. Inbreeding can be used to improve response to selection when h^2 is small by combining inbreeding with between-family selection.

Inbreeding Depression

As inbreeding increases, it often causes a decrease in productivity which is termed "inbreeding depression." Inbreeding depression is a decrease in growth rate, fecundity, etc. that is observed in the inbred group when it is compared to a control population where there is no inbreeding. The severity of inbreeding depression depends on the level of inbreeding, the phenotype in question, and the population. Inbreeding depression is what gives inbreeding its bad reputation.

There are several explanations for why inbreeding depression occurs, and all contribute to inbreeding depression. The first two are qualitative genetic explanations. The most common explanation is that inbreeding depression occurs because of the pairing and expression of detrimental recessive alleles. Although the terms "dominant" and "recessive" do not mean that one allele is good and the other undesirable, most of the alleles that produce abnormal phenotypes or that lower viability are recessive.

Even though mutation rates are usually only one in 10,000 to one in 100,000 replications per gene, because each fish can produce many gametes and because each fish has tens of thousands of genes in its genome, each fish produces dozens to hundreds of gametes that contain a mutant copy of one or more genes. Most of these mutant alleles are recessive. Since recessive alleles can be expressed only when a fish is homozygous, these mutations tend to accumulate in a population. Consequently, many fish carry "hidden" copies of these mutant alleles in the heterozygous state.

Mutations do not have to be bad; some increase fitness or lead to the evolution of new species by creating new or improved phenotypes. But most mutations lower fitness, because they cause random changes in a phenotype that works and that works well as a result of natural selection. Many mutant alleles produce phenotypes that are so abnormal that they either cause death or reduce viability severely. Others cause only small reductions in viability. Each fish carries a number of these detrimental recessive alleles, and inbreeding uncovers them.

When relatives mate, they produce offspring with an increased level of homozygosity; consequently, some of the detrimental recessive alleles that the parents carry in the unexpressed heterozygous state are paired and expressed in the offspring. Detrimental recessive alleles can also be paired and expressed when unrelated fish mate. The difference is one of magnitude, and that is what causes inbreeding depression.

It has been estimated that each individual carries dozens of mutant alleles that lower viability. Most of these mutant alleles will produce only a small reduction in fitness, but several will produce phenotypes so abnormal that they are lethal or cause premature death. There are many of these unexploded genetic bombs in the gene pool. If two unrelated individuals mate, there is a good chance that they will not carry the same genetic dynamite.

However, when relatives mate, there is a good chance that both parents will carry the same genetic time bombs, because they inherited them from a common ancestor. Even though the probability of producing a defective offspring is the same for each recessive allele, if both parents carry a large number of identical defective recessive alleles, the odds of producing defective offspring increase dramatically. The more closely related the parents, the more alike they are genetically, so it is more likely that defective offspring will be created.

Even if two parents do not carry any lethal recessive alleles, they can produce inbred offspring which exhibit inbreeding depression if the parents carry a number of recessive alleles that reduce viability, growth, or other production phenotypes. For example, if two relatives carry 10 recessive alleles, each of which depresses growth by 1%, the pairing and expression of these alleles will depress growth in the inbred offspring. Some offspring produced by these parents will not be homozygous for any of these genes, but most will be homozygous for one or more of these recessive alleles, and growth will be depressed from 1–10% in these offspring.

A second explanation for inbreeding depression is that by decreasing heterozygosity, inbreeding reduces what is called "overdominance." Overdominance occurs when the heterozygous genotype produces a phenotype that is superior to the two homozygous ones. Because inbreeding increases homozygosity, it decreases the number of heterozygous loci, which means that it reduces overdominance. The best known example of overdominance occurs in humans.

In many populations, the gene which produces hemoglobin exists in two allelic forms: one produces normal hemoglobin, while the other produces sickle-cell hemoglobin. Individuals who are homozygous normal are susceptible to malaria and get quite ill and can die when infected. Individuals who are homozygous for sickle-cell get sickle-cell anemia and are sickly and often die. The heterozygotes are superior, even though they are slightly anemic, because they are resistant to malaria.

The final explanation for inbreeding depression is a quantitative genetic one. If a quantitative phenotype were controlled solely by V_A, inbreeding would not effect the phenotype, because the types of homozygosity that would be created at each locus would balance over the genome: some homozygous combinations would cause a slight decrease in the phenotypic mean, while other would cause a slight increase. Since quantitative phenotypes are controlled by hundreds of genes, and since the homozygosity that is created by inbreeding tends to be random, if a quantitative phenotype were solely controlled by V_A, inbreeding-produced homozygosity would neutralize itself, in terms of phenotypic expression. However, few if any quantitative phenotypes are controlled solely by V_A. When V_A controls just 30% of V_P for a particular phenotype, it produces a $h^2 = 0.3$, which is considered to be large.

Since inbreeding increases homozygosity, it also decreases heterozygosity. By doing this, inbreeding affects the interactions between alleles at each locus, which means that it affects V_D, and dominance effects can play a major role in the production of many quantitative phenotypes. When h^2 is small (<0.15), as it often is for viability and fecundity, dominance effects can be quite significant. The creation of excess homozygosity tends to disrupt dominance effects, which tend to be more unidirectional than the additive effects. Inbreeding can be considered to be the opposite of crossbreeding, which is used to exploit V_D. Consequently, if crossbreeding maximizes exploitation of V_D, inbreeding minimizes exploitation, which means that inbreeding makes phenotypes controlled by V_D worse.

This genetic explanation makes sense, because inbred lines are often used in crossbreeding programmes. The creation of inbred lines is done to produce opposite types of homozygosity that will be maximized when hybrids are created, which is how V_D is exploited. This explanation is further supported by the fact that crossbreeding, the breeding programme that is used to exploit V_D, can also be used to eliminate inbreeding and inbreeding depression.

Calculating Individual Inbreeding Values

Calculating individual inbreeding values often requires information that a farmer or hatchery manager does not have-pedigrees. Those who culture fish usually do not know a fish's mother or father. Even if this information is known, it is unlikely that grandparents or cousins can be identified. This information is routinely gathered in livestock husbandry, and family pedigrees are recorded on forms or are entered into computer data bases.

It is far easier to record this information for cattle, pigs, sheep, and goats, because these animals are large, live on land, and can be branded or tagged with little effort at birth. This enables those who work with the marked livestock to identify each individual, to record matings and births, and thus to create pedigrees. Few aquaculturists record such information with fish, because fish are difficult to mark, and marking techniques often kill a large percentage of the fish. Additionally, while this information has been considered important for centuries with livestock, it is still considered to be of little value by most aquaculturists.

Even though it is impossible for most farmers or research scientists to determine individual inbreeding values, it is important to know how these values are calculated and to learn the techniques that are used. An understanding of the methods and protocols needed to determine individual inbreeding values explains how inbreeding is created. It helps explain how an individual can become homozygous by descent, which is what occurs when relatives mate and produce offspring. Learning how to trace family trees and how to calculate individual inbreeding values demonstrates how an ancestral allele can be inherited, both maternally and paternally, from a common ancestor.

An understanding of these techniques also shows how inbreeding can be prevented, demonstrates how inbreeding can be reduced to zero, and helps lay the groundwork for the next chapter, the determination of average inbreeding values in hatchery populations. Two methods can be used to determine individual inbreeding values: path analysis and covariance analysis. Both methods produce the same answer, but some feel more comfortable with one than the other. Both methods use simple arithmetic.

Path analysis is quicker and requires fewer steps, but it is easier to make a mistake using this technique. Covariance analysis can take far more time if the pedigree is complicated, but once complete, the information generated by covariance analysis can be used to quickly predict inbreeding values that will be produced from any possible mating between individuals in the pedigree.

Building a Pedigree

The first step in determining an individual fish's inbreeding is to create its pedigree. A pedigree is a family tree. It lists an individual's brothers and sisters, parents, aunts and uncles, grandparents, great-grandparents, nephews and nieces, and cousins. The information is usually streamlined; often, only direct ancestors are included; and the family tree is often traced back four or fewer generations, only going back to an individual's great-grandparents or occasionally including the great-great-grandparents.

More generations can be listed, but with each additional ancestral generation, the genetic contribution of each fish in the oldest generation to the present generation decreases to the point where it becomes genetically meaningless. This is because an animal's genetic contribution to its descendants is halved each generation: Each parent contributes 50% of a individual's genes; each grandparent contributes 25%; each great-grandparent contributes 12.5%; and each great-great-grandparent contributes 6.25%. The contribution of each great-great-great-great-great-grandparent (seven ancestral generations) is only 0.78%. These values are determined by the number of direct ancestors an individual has in each ancestral generation: two parents, four grandparents, eight great-grandparents, and 16 great-great-grandparents. All fish have 128 great-great-great-great-great-grandparents, which means the contribution from each is so small (0.78%) that it is of no value in most breeding programmes.

Figure 5 contains two pedigrees. Traditionally, males are represented by squares and females are represented by circles. Various symbols can be added to the squares and circles to convey phenotypic information, such as body colour, genetic diseases, etc., but this information is not needed to determine inbreeding. Other information that can be added to a pedigree are: names, birth and death dates, and weights at specific periods. When letters are used to designate individuals, the letter F is usually omitted, because F is the symbol for coefficient of inbreeding; its omission prevents confusion.

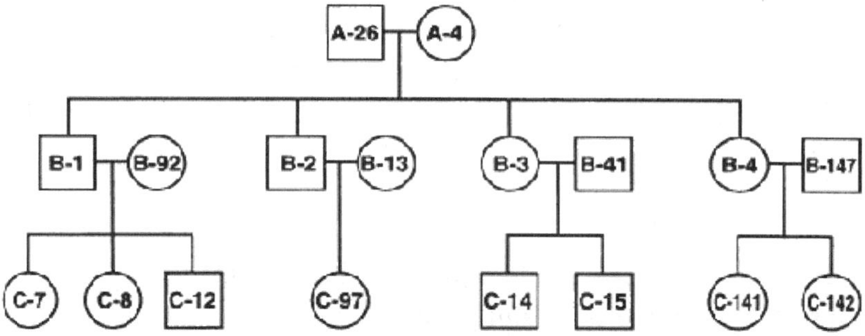

Figure 5. Examples of two pedigrees. The right pedigree uses squares to represent males and circles to represent females. Fish name is inside each square or circle. The other pedigree simply uses letter symbols to represent fish. Information about the fish, such as birth date, weight, etc., would be kept in ledgers or on data sheets. When letters are used to designate individuals, the letter F is usually omitted, because F is the symbol for coefficient of inbreeding; this prevents confusion.

Path Analysis

In path analysis, a pedigree is converted to a path diagram (Figure 6). A path diagram differs from a pedigree in that arrows between parents and offspring replace the tier-step decent brackets that are used in a pedigree. Each arrow represents a gamete, which means that each arrowhead represents the path or way an individual received 50% of its genes, and the arrow shaft touches the parent that contributed those genes. Thus, an arrow shows how a fish received one homologue of each chromosome pair, which means it also shows how a fish received one allele of each allelic pair.

Each fish in a path diagram can be touched by zero, one, or two arrowheads. If a fish's parents are unknown, no arrowhead will touch it; if one parent is known, a single arrowhead will touch the fish; if both parents are known, the fish will be touched by two arrowheads. A fish cannot be touched by more than two arrowheads, because no fish can have more than two biological parents. If this occurs, you have made an error. Many arrows can leave an individual, if that fish produced a number of offspring that are listed in the family tree that is being analysed.

The arrows, or paths, flow from the oldest to the youngest generation, and they show how the genes were transmitted from generation to generation. They do not always flow in a straight line and sometimes cross or go sideways, depending on the complexity of the family tree and the matings that have occurred. Because of this, it is crucial to make the path diagram large enough to house the arrows and to accurately place the arrows. If arrows are incorrectly drawn, inbreeding values generated by the path diagram will be incorrect.

An individual's inbreeding value is calculated by determining all possible paths that the individual has with one or more common ancestors. A common ancestor is an individual that occurs on both sides of the family tree or pedigree; i.e., an individual that contributes genes through both the mother and the father. If just one common ancestor exists, the individual's parents are related and the fish is inbred. If there are no common ancestors, the inbreeding of that fish is zero.

Obviously, if family trees are traced back far enough, all species of fish, like all humans, are related though a common ancestor. However, for practical breeding purposes, common ancestors that are five or six ancestral generations removed are of little importance in any breeding programme, so the genetic contribution of such distant relatives is usually ignored. In most western societies, the legal distance for consanguineous marriages is second cousins; i.e., it is illegal to marry a first cousin or someone more related than that, but is lawful to marry a second cousin or someone less related than that. Western societal norms figure that the inbreeding produced by second cousin marriages is insignificant and therefore acceptable, while that produced by first cousin marriages is not.

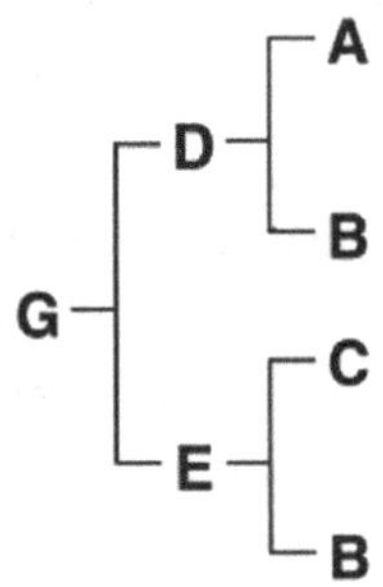

Pedigree

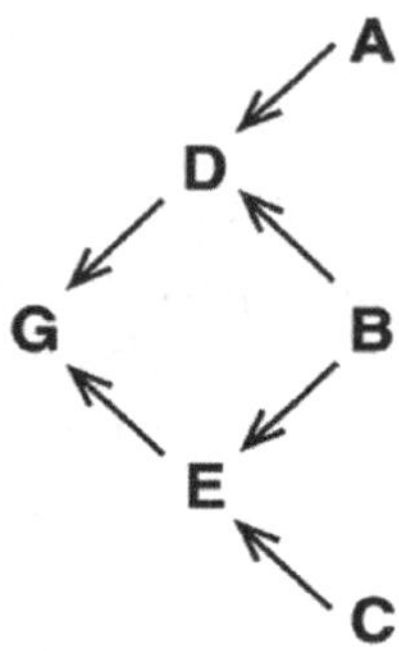

Path Diagram

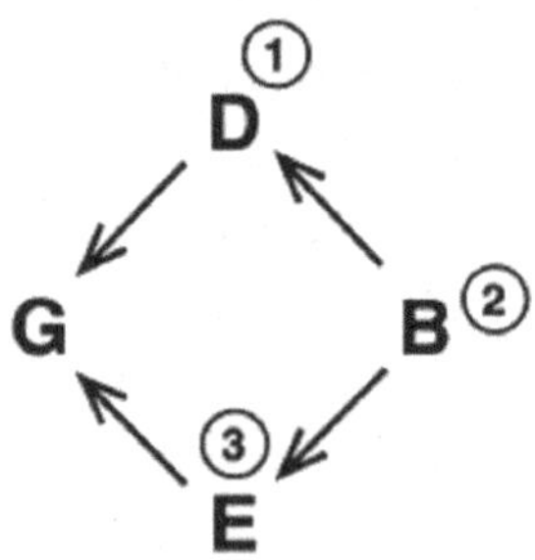

Path from G to common ancestor B

Figure 6. A simple pedigree and the path diagram that can be used to calculate inbreeding. The path that is used to calculate inbreeding of fish G is shown.

The following formula is used to determine individual inbreeding values from a path diagram:

$$F_X = \Sigma[(0.5)^N(1 + F_A)]$$

where: F_X = the inbreeding of an individual;
Σ = the symbol for "sum of " or "add";

N = the number of individuals in a path that is determined by tracing a path from one parent back to the common ancestor and forward from the common ancestor to the other parent; if more than one common ancestor exists, the term $(0.5)^N$ is repeated for each common ancestor; if more than one path exists between the individual and a common ancestor, the term $(0.5)^N$ is repeated for each unique path;

F_A = the inbreeding of the common ancestor.

If the inbreeding value of the common ancestor is zero or unknown, in which case you must assume that it is zero, the formula is simplified to:

$$F_X = \Sigma\,(0.5)^N$$

The formula may look like complicated math, but it is simple arithmetic that can be easily done with basic, inexpensive hand-held calculators; the term $(0.5)^N$ can be determined by using the "y^x" button.

In Figure 6, fish G is the only fish that has a common ancestor. An inspection of the pedigree shows that fish B is on both sides of fish G's family tree; this means that fish G is inbred. No other fish depicted in the pedigree has a common ancestor.

To determine an individual's inbreeding, a path is traced from one parent back to the common ancestor and from the common ancestor up to the other parent. Consequently, fish G's (Figure 6) inbreeding is determined by tracing the path that can be drawn from fish G's parents to fish B; i.e., from fish D (one parent) to fish B (the common ancestor) and from fish B to fish E (the other parent). Therefore, the path that is traced goes:

D-B-E

There are three individuals in this path, so N = 3. Since fish B (the common ancestor) is not inbred, the simplified formula can be used to calculate the inbreeding value of fish G:

$$F_G = (0.5)^3$$
$$F_G = 0.125$$

Thus, $F_G = 0.125$ or 12.5%. This means that fish G is expected to have 12.5% more homozygous loci than the average fish in the population. This also means that 12.5% of the loci that were heterozygous became homozygous as as result of inbreeding.

Figure 7 shows a more complicated pedigree and its path diagram. Fish J and fish N are inbred, because both have common ancestors. Fish A is the common ancestor of fish J. The inbreeding of fish J is determined by tracing a path to fish A through fish J's parents (fish H and fish I). The path is:

H-A-I

There are three individuals in the path, so N = 3. Since fish A (the common ancestor) is not inbred, we can use the simplified formula, and the inbreeding value for fish J is:

$$F_J = (0.5)^3 \; F_J = 0.125$$

Thus, $F_J = 0.125$ or 12.5%.

Fish N has more than one common ancestor. This means that more than one path will determine the inbreeding of fish N. Fish A and fish C are both common ancestors of fish N. A glance at the pedigree reveals that fish A and fish C are the only fish to appear on both sides of fish N's pedigree. Fish J is inbred, but since it does not appear on both sides of fish N's pedigree, its inbreeding value does not contribute to fish N's inbreeding value.

The path that can be traced from fish J to fish M (the parents of fish N) though common ancestor fish C is:

J-I-C-K-M

There are five individuals in this path, so N = 5.

Two different paths can be drawn from fish J to fish M (the parents of fish N) though common ancestor fish A:

1. J-H-A-M
2. J-I-A-M

There are four individuals in each path, so N = 4 for both. Since neither common ancestor is inbred, the simplified formula can be used. The only difference that occurs when there is more than one common ancestor or when more than one path exits for a particular common ancestor is that the inbreeding value is determined by adding the value derived from each path:

$$F_N = (0.5)^5 + (0.5)^4 + (0.5)^4$$

$$F_N = 0.03125 + 0.0625 + 0.0625$$

$$F_N = 0.15625$$

Thus, $F_N = 0.15625$ or 15.62%.

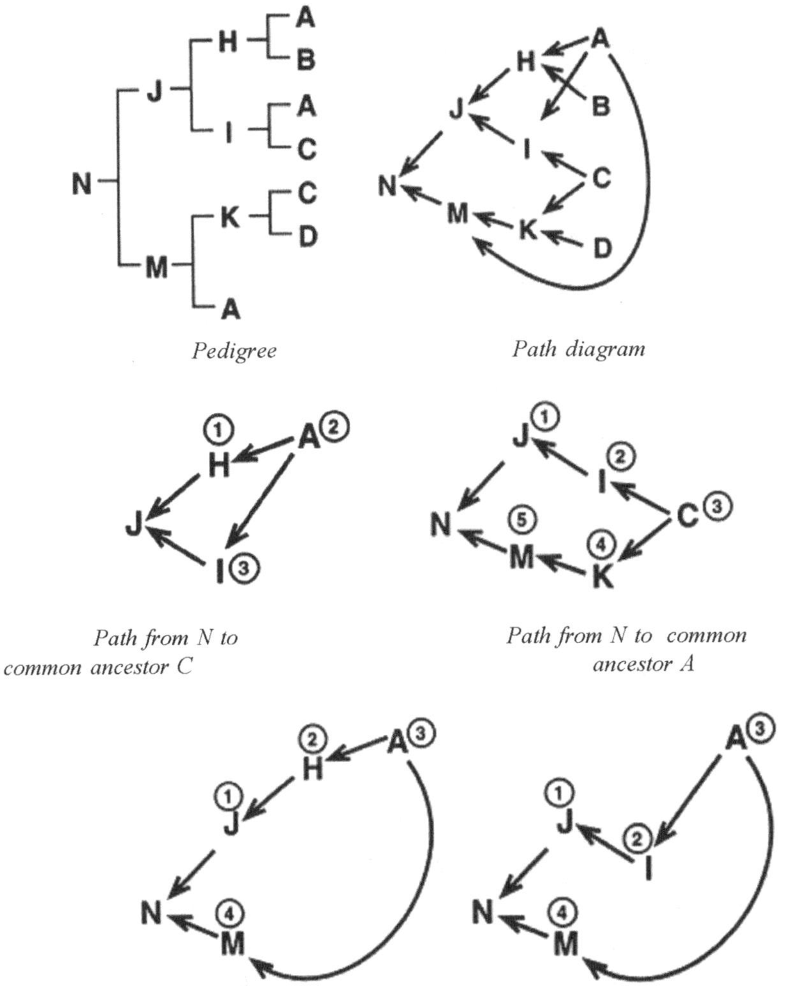

Figure 7. A pedigree and the path diagram that is used to calculate inbreeding. The paths that are used to calculate inbreeding in fish J and fish N are shown

There are two important rules about tracing paths:

Rule 1:

You cannot retrace a path; i.e., you cannot go though any individual twice in a given path. That is why you cannot use the path J-I-C-I-A-M to determine inbreeding of fish N in Figure 7. Fish I occurs twice in that path.

Rule 2:

A path is traced backward from one parent to the common ancestor, and then forward from the common ancestor to the other parent.

The part of the path that goes from one parent to the common ancestor will travel only in a backward direction; i.e., the path it traces will always start at the arrowhead and go towards the shaft. The part of the path that returns from the common ancestor to the other parent will travel only in a forward direction; i.e., the path it traces will always start at an arrow's shaft and go to the arrowhead. A path that travels in both directions, either on the way to the common ancestor or on the way back, is an erroneous path; i.e, two arrowheads or the ends of two shafts cannot touch in a given path. This is why you cannot use the path J-H-A-I-C-K-M when determining the inbreeding of fish N in Figure 7. This path goes forward and backward while going from common ancestor A to parent M; two arrowheads touch when going A-I-C.

Breeding Number and Genetic Drift

Genetic drift is random changes in gene frequency that occur because of sampling error. Sampling error can be natural, or it can be manmade. Natural sampling errors are those which occur when earthquakes, floods, landslides, or other natural disasters subdivide a population and isolate small groups of organisms. This process is a major force in the evolution of new species. Manmade sampling errors are inaccurate collections: sampling only a portion of the population, sampling only a single age class, sampling only fish that possess a certain phenotype or that spawn on a particular day, etc.

When a population is sampled, there is a chance that the sample does not accurately reflect the make-up of the population. This inaccuracy can include length, sex ratio, body colour, and gene frequencies. The smaller the sample, the greater the likelihood that inaccuracies in the sample will occur. Changes in gene frequency that occur as a result of sampling error are called "genetic drift."

When culturing fish, the important changes that can occur in gene frequency as a result of genetic drift occur during the creation of the next generation or during the acquisition of the population. This is when the genes are transferred from parents to offspring or when they are transferred from one hatchery to another. As was the case with inbreeding, N_e is the factor that determines the magnitude of genetic drift. The relationship between N_e and genetic drift is:

$$\sigma^2_{\Delta q} = \frac{pq}{2N_e}$$

where: $\sigma^2\Delta_q$ is the variance in the change of gene frequency, and *p* and *q* are the frequencies of alleles *p* and *q* for a given gene. The variance of the change in gene frequency is the way genetic drift is measured.

Like average inbreeding value, genetic drift is also inversely related to N_e. A comparison of the two formulae shows that they are similar. This means that large N_e's produce small changes in gene frequency, while small N_e's produce large changes in gene frequency.

Many factors can cause changes in gene frequency; among them are: selection by the farmer or hatchery manager; domestication; genetic drift. The changes in gene frequency caused by selection and domestication usually produce genetic improvements, in that the fish grow faster, are more disease resistant, are calmer, accept pelleted feed more readily, have more efficient feed conversions, and are easier to spawn. In contrast, the changes in gene frequency that are caused by genetic drift are random, which means they can be counterproductive. If the frequency of an allele changes from, say, 0.5 to 0.45 or from 0.4 to 0.3 as a result of genetic drift, the genetic effects on the population might not be that great. The major damage that is caused by genetic drift occurs when the frequency of an allele goes to 0.0; i.e., the allele is lost and no longer exists in the population. The odds of losing alleles via genetic drift are related to their frequencies; i.e., rare alleles are lost more easily than common ones.

The probability of losing an allele by genetic drift is determined by using the following formula:

$$P = (1.0 - q)^{2N_e}$$

where: P is the probability of losing an allele, and q is the frequency of the allele.

For example, if N_e is 50, the probability of losing an allele whose frequency is 0.1 (q = 0.1) is:

$$P = (1.0 - 0.1)^{2(50)}$$
$$P = (0.9)^{100}$$
$$P = 0.0000265$$

If N_e is 10, the probability of losing the same allele is:

$$P = (1.0 - 0.1)^{2(10)}$$
$$P = (0.9)^{20}$$
$$P = 0.12158$$

These examples clearly show that the probability of losing an allele is inversely related to N_e; the probability of losing the allele was 4,588 times greater when N_e decreased from 50 to 10.

Table 1 lists the probabilities of losing alleles of various frequencies via genetic drift for N_e's that range from 2 to 6,880. For example, an N_e of 10 produces a probability of losing an allele of P = 0.000001 (1×10^{-6}) for an allele whose frequency is 0.5. This means that there is a virtual guarantee (99.9999% chance) of keeping an allele whose frequency is 0.5 (guarantee of keeping the allele = 1.0 - the probability of losing the allele). On the other hand, the same N_e produces a probability of losing an allele of P = 0.81791 for an allele whose frequency is 0.01; the guarantee of keeping such an allele with an N_e of 10 is only 18.2%. These values demonstrate that rare alleles are more likely to be lost than common ones when N_e is small.

The probabilities listed in Table 1 are for a single generation. Additionally, they are the probability for losing a single allele of the frequencies that are listed. Consequently, if there is a 20% chance of losing an allele of a given frequency, 20% of all alleles with that frequency will probably be lost while 80% will remain, although the frequencies will probably be different as a result of genetic drift. The loss of alleles via genetic drift has two effects: First, it increases homozygosity; consequently, it has an effect similar to that seen for inbreeding. The simultaneous effect of an increase in inbreeding and the loss of alleles via genetic drift as a result of a decrease in N_e can cause severe genetic problems. Secondly, the loss of alleles reduces genetic variance.

Table 1. Probabilities of losing an allele via genetic drift for eight allelic frequencies at various effective breeding numbers (N_e). These probabilities are for a single event. The guarantee of keeping the allele is: 1.0 - the probability of losing the allele. Once the probability of losing an allele reaches 1×10^{-6}, no further probabilities are listed.

	Allelic frequency							
N_e	*0.5*	*0.4*	*0.3*	*0.2*	*0.1*	*0.05*	*0.01*	*0.001*
2	0.06250	0.12960	0.24010	0.40960	0.65610	0.81451	0.96060	0.99601
3	0.01562	0.04666	0.11765	0.26214	0.53144	0.73509	0.94148	0.99402
4	0.00391	0.01680	0.05765	0.16778	0.43047	0.66342	0.92274	0.99203
5	0.00098	0.00605	0.02825	0.10737	0.34868	0.59874	0.90438	0.99004
6	0.00024	0.00218	0.01384	0.06872	0.28243	0.54036	0.88638	0.98807
7	0.00006	0.00078	0.00678	0.04398	0.22877	0.48768	0.86875	0.98609
8	0.00002	0.00028	0.00332	0.02815	0.18530	0.44013	0.85146	0.98412
9	4×10^{-6}	0.00010	0.00163	0.01801	0.15009	0.39721	0.83451	0.98215
10	1×10^{-6}	0.00004	0.00080	0.01153	0.12158	0.35849	0.81791	0.98019
14		6×10^{-7}	0.00005	0.00193	0.05233	0.23783	0.75472	0.97237
15			0.00002	0.00124	0.04239	0.21464	0.73970	0.97043
20			6×10^{-7}	0.00013	0.01478	0.12851	0.66897	0.96077
25				0.00001	0.00515	0.07694	0.60501	0.95121
30				2×10^{-6}	0.00180	0.04607	0.54716	0.94174
31				1×10^{-6}	0.00146	0.04158	0.53627	0.93985
35					0.00063	0.02758	0.49484	0.93236
40					0.00022	0.01652	0.44752	0.92308
45					0.00008	0.00989	0.40473	0.91389
50					0.00003	0.00592	0.36603	0.90479
55					9×10^{-6}	0.00354	0.33103	0.89578
60					3×10^{-6}	0.00212	0.29938	0.88687
66					9×10^{-7}	0.00115	0.26537	0.87628
70						0.00076	0.24487	0.86930
75						0.00046	0.22145	0.86064
80						0.00027	0.20028	0.85208
85						0.00016	0.18113	0.84359
90						0.00010	0.16381	0.83520
95						0.00006	0.14814	0.82688
100						0.00004	0.13398	0.81865
125						3×10^{-6}	0.08106	0.77870
135						1×10^{-6}	0.06630	0.76328
150							0.04904	0.74071
175							0.02967	0.70456

200	0.01795	0.67019
225	0.01086	0.63748
230	0.00982	0.63114
250	0.00657	0.60638
275	0.00398	0.57679
300	0.00240	0.54865
325	0.00146	0.52188
350	0.00088	0.49641
375	0.00053	0.47219
400	0.00032	0.44915
425	0.00019	0.42723
450	0.00012	0.40639
475	0.00007	0.38656
500	0.00004	0.36770
685	1×10^{-6}	0.25393
1498		0.04991
2302		0.00999
6880		$1 \times 10^{-}$

Genetic variance is the raw material with which selection works. A reduction in genetic variance can make selection difficult, if not impractical. If there is no genetic variance, there will be no heritable differences, which means that selection cannot improve a phenotype. Equally important, if the population is being cultured for stocking lakes and rivers, the loss of genetic variance may doom the project to failure. Natural populations need broad gene pools, because it is impossible to predict what genotypes and what alleles will be needed to ensure survival. Populations with narrow genetic bases are less likely to survive in the long term.

The problems caused by genetic drift are measurable, and genetic drift has been shown to have damaged the gene pool of several hatchery populations. Research has shown that brood stock acquisition drastically altered gene frequencies, even when measures were taken to prevent genetic drift. Genetic drift has been shown to have robbed an important aquacultured population of Nile tilapia of so many alleles that there was no detectable heterozygosity and there was virtually no heritable variance for growth. Finally, genetic drift-induced changes in gene frequency may be a major reason why many stocking programmes have been unsuccessful.

References

Bondari, K., and R.A. Dunham. 1986. The effect of inbreeding on reproduction, growth and survival of channel catfish. *Aquaculture* 57:363–364.

Doyle, R.W., and A. Talbot. 1986. Effective population size and selection in variable aquaculture stocks. *Aquaculture* 57:27–35.

Kincaid, H.L. 1983. Inbreeding in fish populations used for aquaculture. *Aquaculture* 33:215–227.

Kirpichnikov, V.S. 1981. *Genetic Bases of Fish Selection.*Springer-Verlag, New York, New York, USA.

Tave, D. 1993. *Genetics for Fish Hatchery Managers,* 2nd ed. Van Nostrand Reinhold, New York, New York, USA.

5

Nutritional Biotechnology in Aquaculture

Globally, consumption of food fishes is projected to reach 165 million tonnes by 2030. As wild catch fisheries are now approaching full exploitation worldwide, a large part of this increase will have to come from aquaculture. The identificcation of alternative fish species and suitable feed ingredients for their diet formulations has therefore become very important.

The main goal of fish nutrition as a scientificc discipline is to produce feeds that support good growth rates while maintaining fish health and quality, resulting in a safe and healthy product for the consumer at least cost. In this regard many scientists are studying the safe utilization of nutrients and their interactions when alternative feed ingredients from plants are used to substitute for traditional (and expensive) fish meal and oil, as well as evaluating alternative marine ingredients. Research focuses on issues such as the characterization of nutrient effects on brooders fish, eggs, larvae, juveniles and at different stages of growth. Measurements include nutrients effect on growth, feed utilization, digestibility, alterations in metabolic pathways, fish health and welfare parameters, nutrient bioavailability and retention. Modern tools within genomics and proteomics are gradually being taken up and focused, giving the opportunity to discover novel pathways and effects of nutrients.

The aquaculture feed sector of India has made tremendous developments during the last two decades. At present about 20 million tones of manufactured aqua feed are being used in aquaculture sector, of which the majority is consumed in shrimp culture. If the rapid growth of aquaculture persists, the feed requirement may increase many fold. Hence, more scientific understanding and interventions are required for sustainable aquaculture of the country.

Sustainable commercial carp feed production has became a challenge to the aquaculture nutritionists. It has been estimated that a feed with FCR 1.3 could make a commercial carp feed sustainable. Exploration of novel genes related to growth enhancement and use of different nutraceuticals have raised the hope of achieving that target. Isolating a potential growth hormone gene and subsequent transfer of that gene to enhance the fish production is approaching near to reality. Addition of new immuno-stimulants in aqua feed has increased the possibility of safer production through high-tech aquaculture. Strategies for increasing utilization of cheaper nutrients like carbohydrate by various technological interventions and addition of feed attractants paved the way for developing lower cost feeds sustainable carp culture. Currently, quality enhancement of fish flesh by dietary and gene manipulation is a focused area of research in fish nutrition.

Besides feed development, feed management and feed quality are critical factors for profitability of fish farming, especially in intensive aquaculture. A biologically ideal feed may not be economically viable if feed management is poor. A balanced knowledge on fish nutrition and feed management is of paramount importance for sustainable fish production. The following are the priority areas of research in fish nutrition:

- Nutritional requirement of fishes with respect to growth.
- Selection of proper feed ingredients.
- Energy requirements in fish.
- Feed formulation and preparation.
- Feed processing and feeding techniques.
- Carbohydrate utilization as protein sparing effect in fish nutrition.
- Omega-3 fatty acids in nutrition and health.
- Nutraceuticals in fish nutrition.
- Fish nutrition, biochemical and physiological responses in extreme environmental stresses.
- Quality control and storage of feed ingredients.
- Anti-nutritional factors, their detoxification and pathology.

- Feed left -over & waste vis-a-vis water quality management.
- Molecular strategies to enhance fish flesh quality and quantity.

Biotechnology is also helping to answer some of the technical and environmental concerns of fish farming including aquaculture nutrition. Many of these centre around what the fish eat. Over the last decade, the world has witnessed spectacular growth in the aquaculture industry of many developing countries. It is further anticipated that world aquaculture production will continue to increase. Since nutrition and feeding play a pivotal role in sustainable aquaculture and since feed constitutes about 40-50% of total cost of aquaculture production. Use of nutritionally balanced and complete formulated feed will continue to play a dominant role in finfish and shellfish production. Hence, alternative and biotechnologically improved feed ingredients should be sought along with improvements in pond management and manipulation of pond productivity.

Right now, the most common protein source for many fish diets is fish meal, a byproduct of fish processing, used because of its high quality protein content. However, it has some disadvantages. In addition to its high cost, it contains levels of phosphorus far beyond the requirement for optimal growth in fish leading to potential environmental concerns. The excess phosphorus released into the water can cause problems such as eutrophication or excess algae growth.

Plant Proteins as Alternatives to Fish Meals

Plant protein has the potential to decrease the problem of phosphorus pollution, since plants do not contain such high phosphorus levels. Moreover, the use of plant protein as a feed ingredient can help to reduce the burden on fish meal supplies. As a result of these concerns, researchers are using biotechnology to produce alternative plant-based protein sources suitable for use in aquaculture.

However, the use of plant-derived materials such as legume seeds, different types of oilseed cake, leaf meals, leaf protein concentrates and root tuber meals as feed ingredients is often limited by the presence of a wide variety of anti-nutritional substances. Plant proteins often require processing to remove these substances, which may be produced by plants as natural defence mechanisms. Important among these are protease inhibitors, phytates, glucosinolates, saponins, tannins, lectins, oligosaccharides and non-starch polysaccharides, phytoestrogens, alkaloids, antigenic compounds, gossypols, cyanogens, mimosine, cyclopropenoid fatty acids, canavanine, antivitamins, and phorbol esters. These compounds must be destroyed during processing to prevent harm to the fish. Problems associated with anti-nutritional factors can also be solved by producing feed enzymes to counteract them. Phytase is one example, which can help fish make the best use of the phosphorous available in a plant-protein based feed. One of the most exciting technological developments has come from the ability to manipulate the plant genome to produce products economically for use in aquaculture. The use of genetically modified crops to eliminate anti-nutritional factors and increase specific nutrients (limiting amino acids, n-3 fatty acids, etc.) is now possible. The packaging of genetically engineered proteins in corn seed to produce very inexpensive oral vaccines is also being pursued.

Feed Additives

Adding specific nutrients to feed can improve animal digestion and thereby reduce feed costs. A lot of feed additives are being currently used and new concepts are continuously developed with the help of biotechnology.

Utilization of Plant Fibres in Fish Feed through Enzymes

Although enzymes have been in use for a long time in other applications such as detergents, textiles, baking and brewing, their use in the animal and fish feed has been a recent phenomenon. This is because enzymes need to be designed to suit the appropriate application. Industrial enzymes are mainly produced from microorganisms by a process of fermentation and extraction. These enzymes can be produced in large quantities from genetically modified microbes with desired properties to make them economically viable. Feed enzymes need to be robust to stand variations in pH and temperatures. They need to have high temperature stability to withstand pelletization and also have a long shelf life. Over the years, feed enzymes have been going through an evolution from liquid to powder to granules in their product forms in order to make them more heat stable. Lately the granulation technology has been developed in a way that the enzyme molecules are coated with an inert material like cellulose and wax to give the enzyme full protection. This technology helps the enzyme to achieve longer shelf life and is most suitable for pelletized feed. In Europe and Australia, feed enzymes have been used for nearly a decade but their usage in Indian fish diets has been only in recent years because the technology for the production of these enzymes was not well developed in India.

Several alternate feed sources like sunflower, rapeseed and safflower can provide protein, and sorghum, millets or rice bran for energy could be used as feed ingredients if supplemented by substrate-specific enzymes. Cell wall composition of these high fiber feed ingredients show that they contain large amounts of arabino-xylans, pectic polysaccharides and some cellulose. Use of specific enzymes like xylanase, pectinase and cellulase could allow breakdown of the fibre releasing energy as well as increasing the protein digestibility due to better accessibility of the protein. In this way the feed cost could be reduced and the protein levels in the feed a increased. The scope for increasing the use of enzymes in the coming decade is certainly there when high quality enzymes become available with more predictable performances.

Phytase in Aquafeed

One of the enzymes that has really caught on in India is phytase, an enzyme that breaks down the indigestible phytic acid (phytate) in cereals and oilseeds and releases digestible phosphorus. It has been known for several decades however the feed industry could not use it economically due to its high production costs. Microbial phytase became commercially available in 1990s as the result of biotechnological improvements. Phytase reduces the use of expensive supplemental inorganic phosphorus such as dicalcium phosphate. Phytase releases minerals (Ca, Mg, Zn and K), amino acids and proteins, which are complexed with the phytate molecule. Phytate itself is an anti-nutritive factor,which when hydrolysed gives better performance in animals.

Today, a substantial numbers of farmers in India are using phytase to reduce the cost of their feed. One interesting observation made from the environmental point of view is in coastal region of Andhra Pradesh. Since phytase has been used in the poultry and aquaculture industries farmers have been benefiting from better control over algal blooms due to the reduction in ground water phosphorus levels. This shows that the phosphorus content in the poultry litter has come down due to usage of phytase which ultimately leads to the reduction of phosphorus in ground water.

Aqua feed manufacturers are also looking into the possibility of using phytase to release the non-available phosphorus from deoiled rice bran, Soya and wheat. There are various enzyme formulations available in the market with varied activities. Because the enzymes used by the feed industry are produced by different microorganisms, the enzyme characteristics as well as the composition of enzymes would be different. The manufacturer by experience and good quality control can guarantee consistent results.

However, comparison of enzyme products is diffi cult except by in vivo testing. Each microorganism produces enzymes with different optimum pH values, different optimum temperature of operation and different affi nity for the substrate in feed. It has not yet been possible to develop an in-vitro method that can predict in-vivo performance. At present, most of the enzyme products in India are directly imported or the individual enzymes are imported and formulated.

These enzymes attract import duty, making their usage expensive. There are very few companies producing enzymes in India due to lack of technology as well as appropriate seed organisms. Most of the multinational enzyme companies have spent years in R & D efforts for development of appropriate enzymes. India needs to look into these aspects and invest in enzyme production biotechnology units in the near future in order to make them economically viable.

Improving expression of the phytase gene in plants is underway as a means to commercially produce phytase, as in biofarming in which plants such as alfalfa are used as "bioreactors" and also by developing plant cultivars that would produce enough transgenic phytase so that their additional supplementation in grain or meals is not necessary.

Micro-nutrients and Vitamins in Fish Feed

The absorption and availability of inorganic trace minerals such as zinc, cobalt, calcium etc., varies depending upon the nature of the minerals (sulphate, oxide or carbonate), their solubility and ionization. Trace minerals are now being attached to oligopeptides to make them more bioavailable. Commercial preparations of proteinated selenium and chromium are used in poultry production. In the case of vitamins, due to varying availability and stability of vitamins in ingredients, supplemental vitamins are widely incorporated in diets. These vitamins are much more stable than naturally occurring forms. Stability is achieved through the application of advanced technologies which involve preparation of biologically active derivatives, coating technologies, carriers and diluents.

Nutraceuticals

Nutraceutical implies that the extract of food is demonstrated to have a physiological benefit or provide protection against a disease and/or improve growth. Functional foods are defi ned as being consumed as part of a usual diet but are demonstrated to have physiological benefits and/or reduce the risk of disease beyond basic nutritional functions. Nutraceuticals are often used in nutrient premixes or nutrient systems in the food. In aquaculture, application of nutraceuticals includes addition of feed additives in feed such as antioxidants,vitamins, minerals, and carotenoids etc. Nutraceuticals are also extracted from many fishes including omega-3 oil, chitosan and glucosamine, originally derived from waste products. Hundreds of tons of marine by-products are available annually which are driving force for both research and commercialization in the area of marine nutraceuticals.

Dietary Amino Acids through GMOs

Essential amino acids are added as supplements to the feed improve the balance of the amino acid profi le. Since the amino acid profiles of feed ingredients do not normally match the profi le of amino acid requirements of the target species, supplemental essential amino acids are added to improve protein utilisation. The new trend is to formulate diets on digestible amino acid levels thereby reducing the requirement for protein. So far lysine and methionine have been used as supplements. Lysine is produced by microbial fermentation and methionine is chemically synthesized.

Genetically enhanced micro-organisms are being used to produce threonine and tryptophan on a commercial basis and soon other essential amino acids also would become available. Using all these amino acids it is possible to lower dietary crude protein level by 2–3 %, which is a substantial saving for the farmer. The concept of an ideal protein blend from GM feedstuffs and feed additives (such as amino acids and enzymes) will greatly help with decreasing the amount of nitrogen excreted in animal waste.

Toxin Binders in Fish Feeds

Feed manufacturers have been incorporating various mold inhibitors in their diets to prevent mycotoxin formation. A variety of physical, chemical and biological approaches to counteract the mycotoxin problem have been reported, but large scale, practical and cost effective means for detoxification of mycotoxins containing feed stuffs are limited. Most of them are fungistats and not fungicides that only inhibit growth of molds and do not inactivate any toxins already present. Present day methods generally use organic acids and their salts like propionic acid or adsorbents like bentonites, zeolites, hydroxyl aluminosilicates. In the future, biotechnology based products like microbes, herbal extracts or esterified glucomannan could be used. Extracts of garlic, onion, turmeric, neem have been shown to exert antifungal activity or inhibit aflatoxin production. Identification of the active ingredients would help in the development of genetically modified herbs with enhanced activities to make them cost effective.

Fish Feed Attractants

Colour, smell, odour, taste and flavour of several food and feed stuffs play an important role in attraction of animals towards a particular feed. New strategies including from the Central Institute for Fisheries Education, Mumbai, are being adopted to isolate, purify and utilize the various active principles and chemical ingredients from indigenous plants and traditional herbs as fish attractants in feed. This may help in reducing feed waste and scheduling of feeding frequency.

Probiotics in Fish Nutrition

Microorganisms are naturally present in the digestive system of the animals. Some microbes aid digestion, others can potentially cause pathogenesis. The microbial ecology of the gut merits greater attention due to its implications for nutrition, feed conversion and disease control. Use of antibiotics disturbs the microbiological balance of gut flora eliminating most of the beneficial flora. On stopping antibiotic treatment, pathogens begin to reestablish themselves in the intestine. Overgrowth of these organisms and subsequent invasion of the system by pathogenic organisms cause inflammatory, immunological, neurological and endocrinological problems. Using probiotics can help build up the beneficial bacteria in the intestine and competitively exclude the pathogenic bacteria. Probiotic bacteria also release enzymes, which help in the digestion of feed. The concept of using probiotics in animal feed particularly poultry and aqua-culture is slowly becoming popular. The common organisms in probiotic products are Aspergillus oryzae, Lactobacillus acidophilus, L. bulgaricus, L.plantarium, Bifi dobacterium bifi dium, Streptococcus lactis and Saccharomyces cerevisiae.

These products can be administered through water or incorporated in the feed. Probiotics have been particularly useful in the early stages of chick growth since the gut of the newly hatched chick is sterile and administering probiotics through water at this stage helps to build up beneficial bacteria much faster than the normal course.

The most important quality parameter of probiotics is that the vegetative or the spore forms have to be viable to be able to multiply in the gut. Secondly they should be resistant to antibiotics, which are administered so that the gut ecology could be maintained. Genetic engineering would help develop probiotics with special properties like secreting enzymes and vitamins in large quantities. Such products would be the future generation feed additives.

Probiotics have also been used in a big way as pond cleaners in aquaculture. Probiotic bacteria directly uptake or decompose the organic matter or toxic material and improve the quality of water. The microbial cultures produce a variety of enzymes like amylase, protease, lipase, xylanase and cellu-lase in high concentrations than the native bacteria, which help in degrading waste. These bacteria have a wide range of tolerance for salinity, temperature, pH which usually exists in aquaculture operations. The use of antibiotics in aquaculture is banned due to rejection of export consignments of marine products. Hence usage of probiotics is propagated to counter the effect of viral and bacterial infections in commercial aquaculture. The pond probiotics also sometimes contain a special blend of denitrifying bacteria that remove the algae's primary source of food, nitrogen from the water. This drastic reduction in nitrogen concentration makes it difficult for the algae to bloom. The balance between phytoplankton, zooplankton and beneficial bacteria during culture period play a crucial role in the maintenance of pond health.

The concept of prebiotics in aquafeed is fairly recent. Prebiotics are basically feed for probiotics that are resistant to attack by endogenous enzymes and hence reach the site for proliferation of gut microflora. Some of the prebiotics that are currently used in animal feed are mannan-oligosaccharides (MOS), fructo-oligosaccharide and mixed oligo-dextran. Mannan-oligosaccharides are mainly obtained from cell walls of yeasts. Other sources of MOS are copra or palm kernel meal. MOS interferes with the colonization of the pathogens. Cell surface carbohydrates are primarily responsible for cell recognition. Bacteria have lectins (glycoprotein) on the cell surface that recognize specific sugars and allow the cell to attach to that sugar. Binding of Salmonella, E.coli and Vibrio sp. is shown to be mediated by a mannose specific lectin like substance present on the bacterial cell surface. Similarly fructo-oligosaccharides from chicory have been used as prebiotics to competitively exclude pathogenic bacteria. The pH of the lumen gets reduced thus preventing the entry of pathogenic bacteria. The concept of using prebiotics has not yet been accepted but the advantages of prebiotics are that it can stand high pelletizing temperatures in the feed and also have a long shelf life.

Management and Detection of Contaminants in Fish Feeds

Modern and biotechnological tools like PCR, biomonitoring and DNA/gene based diagnostics could be developed and used for detection of Aspergillus fungi, Salmonella species, mycotoxins and other contaminants in fish feeds and feeding environment. UV/visible spectrophotometric, HPLC and Ag:Ab based detective technologies could also be utilized for detection of several other health hazardous contaminants, heavy metals, xenobiotics, antibiotics, pesticides and steroids etc., in fish feeds.

Growth Improvement and Enhancement of Quantitative Traits

Enhancement of natural growth rates for fish in aquaculture has been extensively explored, with gains arising from improvements in husbandry, nutrition and genetic selection.. Growth enhancement can provide advantages for aquaculture by shortening production times, enhancing feed conversion efficiency, and controlling product availability. Endocrine approaches to controlling growth have also been extensively explored, principally applications of somatotropins such as growth hormone (GH), prolactin, and placental lactogen, insulin- like growth factor-1, thyroid hormones and sex steroids. In the recent past, there are instances both in India and abroad where growths of fish have been increased by incorporation of growth hormone gene with strong promoters.

Highly Unsaturated Fatty Acids (HUFA) and Qualitative Enhancement

Highly unsaturated fatty acids (HUFA) are receiving considerable attention due to their involvement in human health. Fish is an important dietary source of long chain C20 and C22 highly unsaturated fatty acids (HUFA), which are crucial to the health of higher vertebrates also. They play pivotal roles in number of biological functions including cardiovascular functions, neural development, eicosanoid signaling, regulation of gene expression and ion channel modulation.

Traditional emphasis has been on dietary lipids and oils because of their effects on lipid composition, in particular that of the lipoproteins. The classical way by which lipids were evaluated, however, gradually changed in the last decade as it became evident that fatty acids themselves regulate lipid homeostasis not only at the level of the lipids interacting with proteins, but also on the genetic level by affecting gene expression. Changes in the lipid content at high concentration are the major cause for several diseases, i.e., obesity, coronary heart disease and cancer. Earlier studies clearly showed that risk of coronary heart disease was greatly reduced in populations where fish is major portion of food.

In the recent years research work has been initiated to conduct molecular analysis and characterization of delta-6 fatty acyl desaturase genes from several fin fishes which are involved in biosynthesis of omega-3 and omega-6 PUFA.

Genetic Manipulation in Bacterial Biofilms

Molecular approaches to characterizing biofilm structure and development offer considerable potential for finding novel biofouling and disease prevention strategies in enhancing aquatic productivity. It is now possible to determine the genes and pathways involved in regulation and synthesis of bacterial adhesive polymers and virulence antigens. Considerable progress has been made in understanding the nature and expression of surface polymers and bacterial vaccines produced by microorganisms such as the nitrogen-fi xing Rhizobium species and the opportunistic pathogen Pseudomonas aeruginosa etc. Similar approaches can be applied to marine biofilm bacteria, to find the genetic determinants of adhesive production and the environmental factors that regulate synthesis.

Biomaterials also hold promise for counteracting biofouling, which long has been recognized as an extensive and costly problem. Bacterial biofilms form slime layers that increase drag on moving ships, interfere with transfer on heat exchangers, block pipelines, and contribute to corrosion on metal surfaces. Bacterial and microalgal colonization of surfaces is accompanied by settlement of invertebrate larvae and algal spores, eventually leading to "hard fouling" and the need for costly cleaning. The most effective antifouling coatings have utilized toxic chemicals, such as copper and organotins. There is an urgent need for non-toxic biofouling control strategies, due to heightened recognition of the impact that toxic coatings can have on the environment. Research is needed on the attachment mechanisms of marine organisms and the natural products they employ to prevent fouling of their own surfaces.

Developments of Biosensors

Marine organisms can provide the basis for development of biosensors, bio-indicators, and diagnostic devices for medicine, aquaculture, and environmental monitoring. One type of biosensor employs the enzymes responsible for bioluminescence. The lux genes, which encode these enzymes, have been cloned from marine bacteria such as Vibrio fischeri and transferred successfully to a variety of plants and other bacteria. The lux genes typically are inserted into a gene sequence, or operon, that is functional only when stimulated by a defined environmental feature. The enzymes responsible for toluene degradation, for example, are synthesized only in the presence of toluene. When lux genes are inserted into a toluene operon, the engineered bacterium glows yellow-green in the presence of toluene. This genetically engineered system "reports" that biodegradation of a specific chemical, in this case toluene, is proceeding. Another type of biomonitor that holds great promise is the gene probe, which can be used to identify organisms that pose health hazards or may be useful in research. Specific gene probes can be employed, for example, to detect human pathogens in seafood and recreational waters; fish pathogens in aquaculture systems; microorganisms capable of mediating desired chemical transformations (e.g., toxic chemical degradation, CO_2 assimilation, metal reduction); and specific fish stocks in fish migration and recruitment studies.

Bioremediation for Improving Aquaculture

Bioremediation shows great promise for addressing problems in aquatic environments wherein fishes are allowed to grow and fed with formulated and natural feeds in aquaculture. These problems include catastrophic spills of oil in harbors and shipping lanes and around oil platforms; movement of toxic chemicals from land, through estuaries, into the coastal oceans; disposal of sewage sludge in rivers/water bodies, bilge waste, and chemical process wastes; reclamation of minerals, such as manganese; and management of aquacul-ture and seafood processing waste.

Nutrigenomics

Nutrigenomics is the study of molecular relationships between nutrition and the response of genes, with the aim of extrapolating how such subtle changes can affect animal health. Nutrigenomics focuses on the effect of nutrients on the genome, proteome, and metabolome. It is applying the sciences of genomics, transcriptomics, proteomics and metabolomics to animal nutrition in order to understand the relationship between nutrition and health. Nutrigenomics has also been defined as the application of high-throughput genomic tools in nutrition research. This refers to genetic tools that enable literally millions of genetic screening tests to be conducted at a single time. When such high throughput screening is applied in nutrition research, it allows the examination of how nutrients affect the thousands of genes present in the animal genome. Nutrig-enomics involves the characterization of gene products and the physiological function and interactions of these products. This includes how nutrients impact on the production and action of specific gene products and how these proteins in turn affect the response to nutrients. Nutrigenomics has been associated with the idea of personalized nutrition based on genotype. While there is hope that nutrigenomics will ultimately enable such personalised dietary advice, it is a science still in its infancy. Integration of genetic and metabolic studies using the zebrafish as a model affords an opportunity to dissect the fundamental pathways coordinating growth, nutrition and energy homeostasis. Goldsmith et al. revealed a critical transition in growth control during zebrafish caudal fin development.

Marine and Aquatic Biomass for Energy Production

Approximately 40 percent of all primary energy production, or photosynthesis, occurs in the seas. In this process, oceanic plants (phytoplankton, seaweeds, seagrasses) take up carbon dioxide and convert it into organic carbon (primarily sugars) and oxygen using light from the sun as an energy source. The oceans contain 50 times as much carbon dioxide as does the atmosphere, and it is estimated that primary production incorporates 35 gigatons of carbon into marine biomass annually. This abundant source of fuel for energy production has not been tapped commercially because it is not competitive with soybean meal and other easily harvested, traditional sources of biomass, and also because biomass is not presently competitive with other types of fuels, regardless of its source.

Among three strategic technologies being adopted in abroad, the first one , the enzyme that captures CO_2 for photosyn-thesis—ribulose bisphosphate carboxylase\oxygenase or "RUBISCO"—is relatively inefficient, so supercomputers are being used to verify structural information, and the enzyme is being redesigned to optimize its function. Second, the chemical composition of biomass can be altered to make it more suitable for particular applications. For example, marine microalgae are being genetically engineered to boost their lipid content, with the aim of providing a source of alternative fuels that is more economical than are conventional sources. Third, biotechnology is being used to convert biomass to ethanol and other alternative forms of energy and chemical feedstocks.

Sustainable Marine Resources for Organic Aquafeeds

Expanding organic production rules from terrestrial livestock to fish and other aquatic animals produced by aquaculture is complicated by ecological reality pertaining to food sources available to land animals and birds compared to sources available to fish. On land, plants use sunlight to convert CO_2, nitrogen and other soil nutrients into plant tissue that animals eat, e.g., leaves and seeds. The primary form of stored energy in plants is starch, a polymer of glucose. In the aquatic environment, algae and aquatic macrophytes also use sunlight to convert CO_2, nitrogen and other nutrients into algae/plant tissue but there is a major difference in how energy is stored in that starch is not the primary form of stored energy, and there are no seeds (grains) in the aquatic environment at the base of the animal (fish) food chain. Put simply, algae are consumed by zooplankton which are, in turn, consumed by pelagic fish. The pelagic fish are the food source for so-called carnivorous fish, but technically speaking, the pelagic fish are themselves carnivores in that they consume zooplankton rather than algae. This fundamental difference between the terrestrial and aquatic food chains complicates the direct application of organic rules pertaining to ingredients used in livestock feeds to some important ingredients used in aquafeeds, specifically fish meal and fish oil derived from pelagic fish. This difference also complicates the distinction made in terrestrial food webs between herbivorous, omnivorous and carnivorous animal species when it is applied to the aquatic food web and various trophic levels. Essentially, all fish species, with very few exceptions, are carnivores; although a more accurate term would be piscivorous (primary food source in nature is other fish). Even species of fish commonly thought to be omnivorous, such as channel catfish, tilapia and carp, require high protein feeds at first feeding through the fingerling stage because their digestive tract is not fully developed at first feeding. In nature, these fish consume zooplankton as their primary food source. In aquaculture production, these species are fed fish meal-based feeds until they reach the stage of development where they can be converted to grain and oilseed-based feeds. Contrary to the assertions of some that there would be no need to use fish meal in aquafeeds if only catfish, tilapia and carp were produced by aquaculture rather than salmon, trout, cod, halibut, red sea bream, European sea bass and other piscivorous fish species, the reality is that if fish meal were eliminated from all aquafeeds, production of nearly all species of fish would be difficult if not impossible. At our current level of knowledge in the area of fish nutrition and aquafeeds, fish meal is an essential component of aquafeeds to promote normal growth and health of most species of fish. Growth performance and feed efficiency is reduced in most species of farmed fish fed aquafeeds in which fish meal is completely replaced with plant proteins. This observation is not limited to piscivorous species but extends to omnivorous species at the fry and fingerling stages of growth.

Another aspect of the aquatic food web compared to the terrestrial food web is the presence of highly unsaturated fatty acids (HUFAs) of the omega-3 family in the aquatic food web and the absence of HUFAs in leaves and seeds of plants. Leaves and seeds do contain omega-3 fatty acids, but they are limited to shorter chain versions, specifically linolenic acid (C18:3 n-3) and stearidonic acid (C18:4 n-3). Some species of fish have a limited capacity to convert linolenic and/or stearidonic acid to HUFAs, but most species, notably marine fish, cannot convert the omega-3 fatty acids found in plant oils to HUFAs at a rate essential for normal growth and health.

Omega-3 fatty acids are important nutrients in the human diet and essential nutrients in the diet of fish. Animals, birds, humans and fish cannot synthesize omega-3 fatty acids, but obtain them in their diet. Marine and freshwater algae synthesize omega-3 fatty acids and are the source of all long-chain omega-3 fatty acids in aquatic food webs. Consumers obtain most of their long-chain omega-3 fatty acid intake from fish. Most animals including fish exhibit fatty acid profiles that reflect their diet. Fish fed feeds containing primarily plant oils will have fatty acid profiles in their tissues that resemble plant oils, whereas fish fed feeds containing fish oil have tissue fatty acid profiles will resembling fish oil. Fish oils are therefore essential dietary ingredients in aquafeeds to supply the dietary requirement for fatty acids of the omega-3 family, preferably as HUFAs, and also essential ingredients in aquafeeds to ensure that fillets destined for human consumption contain healthful levels of omega-3 fatty acids expected by consumers, especially those consuming organic products. Fish also require a dietary source of HUFAs to prevent deficiency signs; HUFAs are essential dietary nutrients for fish. However, between 1% and 2% HUFAs in aquafeed is sufficient to prevent essential fatty acid deficiency signs in farmed fish, and this level of HUFAs supplementation is insufficient to ensure adequate HUFA levels in fillets for the consumer. A dietary level of at least 12% over the production cycle is needed in salmon and trout to achieve desired levels of HUFAs in fillets.

The importance of fish meal and fish oil in aquafeeds is not in doubt and aquafeeds used in non-organic aquaculture production will continue to contain both fish meal and fish oil. Strict application of organic rules developed for crop or livestock production to stocks of fish traditionally used to produce fish meal and to aquaculture production will likely result in severe limits on the percentage of fish meal and oil allowed in organic aquafeeds in the USA. Such an outcome will simply illustrate the inappropriateness of applying rules developed for terrestrial animal production to aquatic animal production. Interestingly, other countries, notably the UK, have faced the same problems and found ways to allow fish meal and oil into organic aquafeeds without abandoning the principles of organic production. They have done so by recognizing inherent differences between the terrestrial and aquatic ecosystems, and by applying organic principles in ecosystem-appropriate ways to deal with the main issues associated with production and use of fish meal and fish oil in aquafeeds, namely sustainability, traceability, and quality. In addition to the UK, most international organic standards allow the use of meal and oil produced from fish processing by-products in organic aquafeeds.

Sustainability of Wild Stocks Used to Produce Fish Meal and Fish Oil

Sustainability has many interpretations, but in fisheries, sustainability refers to the way in which harvests of wild stocks are managed to maintain stock abundance at healthy, sustained levels. Stocks of fish are subject to a variety of environmental factors that determine spawning and year-class recruitment success, and large variations in stock abundance unrelated to harvest pressure are well documented. Thus, stocks must be managed conservatively to avoid compounding the effects of variable recruitment success with excessive harvest, thereby reducing stock abundance. Further, harvest management must consider recruitment into spawning populations, and allow sufficient escapement of maturing fish to perpetuate the stock. In the Americas, two species of pelagic fish dominate the production of fish meal and fish oil, Peruvian anchovy *(Engraulis ringens)* and gulf menhaden *(Brevoortia patronus)*. Both stocks are managed to maintain sustainability by restricting harvest, and both have relatively long histories of stock assessment to document population abundance and variability in year-class strength . Landings of the Peruvian anchovy are the highest of any species of fish in the world, ranging from 11, 276,000 metric tons (mt) in 2000 to 6,204,000 mt in 2003, an El Niño year. Because the Peruvian anchovy is a short-lived fish, maturing at about one year of age and having an average longevity of three years, the effects of El Niño are also short-lived; stock abundance typically exceeds average values in the year following an El Niño event. Further, extensive research on Peruvian anchovy stocks was conducted in the 1990s, leading to the establishment of strict harvest quotas and modified fishing techniques to reduce harvest of juvenile fish. Illegal fishing is minimized by aggressive surveillance methods and, as a result, stock biomass has remained healthy. Harvest rates of Peruvian anchovies have been adjusted to prevent overharvesting during fishing seasons when fish are less abundant. Menhaden stocks are also harvested well below their biological capacity and controls on the fishing are comprehensive.

There are several other significant sources of fish meal and fish oil in the Western Hemisphere associated with processing by-products from sustainably managed fisheries, specifically the Alaska pollock (*Theragra chalcogramma*) fisheries in the Bering Sea and Pacific salmon fisheries in Alaska (*Oncorhynchus* sp.). These fisheries have been certified as being sustainably managed by the Marine Stewardship Council . The Alaska pollock fishery is considered the best managed fishery in the world and is second in landings only to the Peruvian anchovy in terms of rankings of capture fisheries in the world. NOAA Fisheries, formerly the National Marine Fisheries Service, is the federal agency responsible for managing the nation's living marine resources. NOAA Fisheries conducts annual trawl and hydro-acoustic surveys to assess Alaska pollock stock abundance and determine the Acceptable Biological Catch level. This value is used to calculate the Total Allowable Catch level which is then set at or below the Acceptable Biological Catch level . In addition, each fishing vessel is assigned two NOAA Fisheries observers who sample the nets to record by-catch. Records indicated that by-catch is minimal, approximately 1% and most of the by-catch is retained and processed, making the discard rate approximately 0.5%, a fraction of the average discard rate of 25% for the world's fisheries. Total landings of Alaska pollock were 1,414,962 mt in 2005 . Of this total, 66% was processing by-product, e.g., material remaining after filleting comprised of the head, viscera, backbone with retained muscle tissue, and skin. Fish meal and fish oil are produced from by-product material in shore-based fish processing plants and onboard fishing vessels using conventional wet reduction processing methods. Since the material processed to make fish meal and oil is removed from human-grade pollock or salmon, the quality of pollock or salmon fish meal or oil is very high.

Five species of Pacific salmon are native to Alaska, and collectively they represent the second largest fisheries in Alaska after pollock. Total salmon harvest in 2005 was 408,014 mt, resulting in 110,164 mt of by-product. Production of salmon meal and oil is a seasonal activity limited to one processing plant, Kodiak Reduction, located in Kodiak Alaska. The biggest consideration concerning the production of organic fish meal and oil in Alaska from fish processing by-products will be the necessity of having strict limits on inclusion of other species in batches of processing by-product. Very small numbers of non-targeted species, such as skates and rays, are always captured in the pollock fishery, and either discarded at sea, or added to processing by-product used to make fish meal and oil. Strict limitations on allowable amounts of by-catch in organically labeled fish meal and oil produced from fisheries processing by-products must be established. Salmon, in contrast, are harvested in targeted fisheries that do not result in capture of non-targeted species. The primary challenge to producing pure salmon meal is concurrent processing of non-salmonid fish during the salmon fishing season, and the necessity of keeping by-products from fish processing plants in Kodiak that are handling other species of fish from being combined with salmon processing by-products.

Traceability of Fish Meal and Fish Oil Produced from Wild Stocks

Traceability refers to product identification along the chain of custody from production to the end user, and is a critical element of concern with regard to the possibility of producing fish meal that is labeled organic from processing by-products or whole fish in the case of menhaden or anchovy meal. Conventional fish meal contains an added antioxidant, ethoxyquin, that is not allowed under organic rules. Hence, any fish meal or oil destined for organic certification must be made in a separate batch from conventional fish meal or oil to avoid contamination with ethoxyquin, and to allow addition of natural antioxidants permitted in organic production. Product traceability is essential to maintain product identification, but maintaining traceability is a minor issue for fish meal and fish oil producers, who routinely produce different grades of fish meal for different markets and therefore must maintain product identity in fish meal and oil production plants in Alaska.

Utilizing processing waste from organically-reared fish to produce fish meal and oil for organic aquafeeds is also feasible. The main priority in utilizing this potential resource will be to avoid recycling of fish pathogens. Therefore fishmeal and fish oil derived from processing by-products of organically-farmed fish must be subjected to appropriate thermal treatment and should ideally not be used in aquafeeds used to feed the same or closely related farmed species of fish. Unfortunately, fish meal and oil from organically-reared fish is not available in the USA because standards for producing organically-reared fish have not yet been adopted.

Quality of Fish Meal and Fish Oil Produced from Wild Stocks

Quality of fish meal and fish oil refers to degree of freshness of the starting material, degree of lipid oxidation, and levels of environmental contaminants, primarily methyl mercury and organic compounds such as PCBs, PBBs, dioxin and related compounds, collectively known as persistent organic pollutants or POPs. Organic rules related to production of grains and oilseeds used to produce feed ingredients are concerned with exposure of crops to POPs, and rules are designed to ensure that crops are not exposed. In the case of fish, no such oversight can be made because wild-caught fish swim freely in the ocean and theoretically may be exposed to persistent organic pollutants (POPs) prior to being harvested. Nevertheless, the US Organic Food Production Act specifically allows organic labeling of wild products, as does the Final Rule. However, analytical testing can detect minute amounts of POPs in fish meal and oil, making it possible to test for the presence of POPs in fish meal and oil, and select products of consistently high quality for use in organic aquafeeds. Further, new processes to remove POPs from fish meal and oil have been adopted in the EU and in the menhaden fish meal and oil industry in the USA that lower POP levels to miniscule amounts . These processes involved removal with activated carbon and can be done economically on large batches of fish meal and oil. This is an important point because POP levels in farmed fish are directly related to POP levels in their feed, nearly all of which are associated with fish meal and fish oil in the feed. Farmed fish consuming feeds containing fish meal or oil with elevated POP levels would accumulate these in their tissues. Fortunately, fish meal and fish oil produced from Peruvian anchovies, menhaden or Alaska pollock/salmon processing by-products has very low POP contamination. Elevated POP levels in fish meal and fish oil have been noted in some products produced from Baltic or North Sea pelagic fish, but treatment with activated carbon reduces POP levels in fish meal and fish oil made from these sources to levels below EU regulated levels.

The level of POPs found in fishmeals and fish oils is very related to the location of the fishery. Concerning levels of methyl mercury in fish meal and fish oil, levels in relatively short-lived fish are orders of magnitude below levels found in long-lived fish at the top of the aquatic food chain, such as marlin, swordfish or large sharks. Levels of methyl mercury are very low in farmed fish, and have never been a concern because the ingredients used in aquafeeds, including fish meal and fish oil, contain negligible levels of methyl mercury.

In regard to degree of freshness and degree of oxidation of the raw material used to make fish meal and oil, and of the final products, fish meal and oil produced from Alaska pollock are very fresh and stable, in part because they are made from human-grade material and in part because they are caught in cold water and held prior to processing in refrigerated sea water. Menhaden and anchovy, in contrast, are captured in warm water and sometimes held without refrigeration prior to processing. Clearly, strict standards must be developed for menhaden and anchovy meals to ensure that they are produced from fresh starting material. Fortunately, degree of freshness is easy to monitor using chemical tests, as is degree of lipid oxidation. Standards for organic fish meal and oil from these species can be developed quickly as the knowledge base in this area is solid.

Need for Fish Oil in Aquafeeds

A final issue concerning the use of fish meal and oil in organic aquafeeds is the nutritional quality of farmed fish for the consumer, specifically their content of omega-3 fatty acids. As mentioned earlier, the fatty acid profiles of fish, livestock and even humans reflect dietary fatty acid intake. Farmed fish fed diets containing primarily plant oils have fatty acid profiles that are similar to plant oils. Fillets from such fish are not nutritionally comparable to fillets from fish fed feeds containing fish oil, the primary source of long chain omega-3 fatty acids. The presence at slaughter of sufficient quantities of eicosapentanoic acid (EPA) and docosahexanoic acid (DHA) in fish fed diets containing fish meal and oil confers clear and well documented health advantages to the consumer. There is no evidence that the presence of omega-6 or short chain omega-3 fatty acids, which would result using feed containing plant oils, confers such health advantages – rather the opposite. In the USA, the omega-6:omega-3 ratio in the average Western diet is 17:1, rather than 2:1 as recommended by health authorities. The only way this ratio can be lowered is for consumers to reduce their intake of corn and soybean oil (and products containing them), both high in omega-6 fatty acids, and increase their intake of fish oil, either directly as a supplement or indirectly through consumption of high-fat fish fillets, such as salmon or trout. Consuming wild salmon is an ideal approach for consumers to increase their intake of HUFAs, but unfortunately, there is not enough wild salmon to provide the quantity in demand in USA markets. Total salmon harvest in Alaska was 408,014 mt in 2005. Assuming a recovery of fillets of approximately 50% of the round weight of salmon, Alaska produced 204,000 mt, or 204,000,000 kg of edible salmon fillets but nearly 65% of Alaska harvest was pink salmon, nearly all of which is canned. Per capita consumption of salmon in the USA increased from 0.72 kg in 2000 to 1.105 kg in 2005, and at a current USA population of 350 million, total salmon consumption in the USA is approximately 387 million kg, well above the quantity of Alaska's salmon harvest. Currently, farmed salmon production is approximately 2.5 times that of wild salmon harvest.

With the exception of grass carp, rabbitfish and a few other species of fish that consume vegetation, virtually all species of farmed fish are carnivorous during some stage of their life history, in contrast to livestock and poultry. This is likely the result of fundamental differences between the terrestrial and aquatic ecosystems, primarily in how energy from the sun is captured and converted to stored energy (starch) in seeds in terrestrial compared to aquatic plants, where energy is stored as fatty acids (lipid). Fish meal is a valuable and, in many cases, an essential ingredient in feeds for many farmed fish species, and limiting or forbidding its use in aquafeeds violates the organic principle of providing feeds that promote well-being and good health in farmed animals and fish. Although scientists have made significant progress in understanding the nutritional requirements of farmed fish, and are using this knowledge to formulate aquafeeds containing much lower levels of fish meal than was the case a decade ago, many species of farmed fish important to the aquaculture industry and to the consumer cannot be produced yet using feeds lacking fish meal. Fortunately, there are sustainable fisheries resources such as Peruvian anchovy and Gulf of Mexico menhaden that, with suitable standards in place, could supply fish meal to the organic aquafeed industry. Further, the Alaska pollock and salmon fisheries, already certified as sustainably managed by the MSC, are poised to produce fish meal and fish oil to organic standards from human-grade fish processing by-products. Given that this material is from sustainably managed fisheries and easily recovered and produced following organic standards for quality and traceability, and also conforms to organic principles of utilizing byproducts if possible, it is logical to utilise this high-quality resource as a source of fish meal and fish oil for organic aquafeeds.

References

Hardy, R.W. 2006. Removing persistent organic pollutants from fish meal and oil. *Aquaculture Magazine* 32(1): 43-45.

Gatesoupe, F. J., 1999. The use of probiotics in aquaculture: a review. *Aquaculture* 180: 147-165.

Gopakumar, K. 2003. Indian aquaculture. *Journal of Applied Aquaculture,* 13(1/2): 1-10.

Muller M., Kersten, S., 2003. Nutrigenomics: Goals and Perspectives. *NatureReviews Genetics* 4. 315 -322.

6

Biotechnology for Aquaculture Development

Aquaculture, which long has been practised in Asia and is increasingly popular in the United States, Europe, and South America, will benefit tremendously from the use of new molecular tools and processes. With worldwide seafood demand projected to increase 70 percent in the next 35 years, and harvests from capture fisheries stable or declining, aquaculture will have to produce seven times as much seafood as it generates now to supply global demand by 2025.

The use of modern biotechnology to intervene in the rearing process and enhance production of aquatic species holds great potential not only to meet this demand, but also to improve competitiveness in aquaculture. Biotechnology is used in several different ways in aquaculture. Researchers use transgenics to introduce desirable genetic traits into the fish, thereby creating hardier stock. Transgenics involves the transfer of genes from one species into another species, in this case, fish. Biotechnology is also applied to research new feed sources and to improve the composition of the feed. This is where much of today's aquaculture research is focused. Other ways in which biotechnology is applied to aquaculture include: improvement of growth rates and control of reproductive cycles through hormone therapy, production of new vaccines and development of disease resistance in fish.

The growth and international competitiveness of aquaculture industry will be determined by the size of the resource investment in research and technology development. This investment should be made through a partnership of central and state agencies and the private sector. The government's role is to provide leadership in supporting research to advance knowledge in important research areas and to facilitate the transfer of promising results and technologies to the private sector.

The major research issues in aquaculture are similar to those for other agricultural sectors, but the knowledge base for aquaculture is comparatively meager. Development of this knowledge is a particular challenge due to the diversity of cultured aquatic species and the systems for their production.

The application of biotechnology promises significant benefits to both producers and consumers of aquacultural products. The use of genetically enhanced organisms may improve production efficiency through improvements in growth rates, food conversion, disease resistance, and product quality and composition. The application of biotechnology to aquaculture also may help conserve wild species and genetic resources and provide unique models for biomedical research.

Oceanic organisms are of enormous scientific interest, for two major reasons. First, they constitute a major share of the Earth's biological resources. Second, marine organisms often possess unique structures, metabolic pathways, reproductive systems, and sensory and defense mechanisms because they have adapted to extreme environments ranging from the cold polar seas at -2°C to the great pressures of the ocean floor, where hydrothermal fluids spew forth. Most major classes of the Earth's organisms are primarily or exclusively marine, so the oceans represent a source of unique genetic information. The oceans therefore offer abundant resources for research and development. Yet the potential of this domain as the basis for new biotechnologies remains largely unexplored. Indeed, the vast majority of marine organisms have yet to be identified. Even for known organisms, there is insufficient knowledge to permit their intelligent management and application.

Interest in marine biotechnology has been growing in recent years, and recent developments in molecular biology promise fundamentally new approaches and opportunities for identifying, using, and managing biological resources from the seas and other bodies of water. These resources have broad potential as technologies for application in many fields, including health, agriculture, bioremediation, and manufacturing. These technologies also will enhance understanding of the basic biology of the seas and their pivotal role in preserving the global environment. With their vast genetic and physiological diversity, marine organisms will be sources for new classes of chemicals and processes, including pharmaceuticals, polymers, enzymes, vaccines, and diagnostic and analytical reagents. New and prosperous industries will be created; one growing industry that will benefit from these developments is aquaculture — the farming of aquatic organisms, including fishes, mollusks, crustaceans, and plants, for food and other products.

In recognition of the potential of the seas to contribute to economic growth and development of both industrialized and developing nations, marine biotechnology institutes have been established during the past decade in many countries, including the United States, Japan, Germany, and India.

Improved Products from the Seas

Most major groups of living organisms are primarily or exclusively marine. Tropical marine environments harbour an especially wide diversity of animals and plants. Many marine organisms are sessile and must employ sophisticated methods to compete for a place to anchor. This characteristic is reflected in part by a metabolism that produces enormously diverse bioactive products — many with no terrestrial counterparts. Recent research has uncovered unicellular and multicellular microorganisms that are unique to the marine world. Indeed, "... it seems clear that marine bacteria are emerging as a significant chemical resource". Before this vast biological resource can be developed and exploited, however, scientists must find answers to fundamental questions concerning the nutritional requirements of marine microorganisms and the amounts and characteristics of chemicals they produce.

Government support for research in this area is essential, both to answer the fundamental questions and to assure that the resulting knowledge is translated into sustainable technologies. While it will be vital to cultivate marine microorganisms that produce novel products, the alternative approach of transferring genes of interest into non-marine microorganisms also should be investigated. For example, the capability to produce a marine polysaccharide — a complex molecule that could be useful as a food additive or a water-resistant adhesive — could be transferred to an easily grown bacterium (e.g., E. coli or Bacillus subtilis). This approach might be more effective in some cases than would cultivating the marine organism or recreating artificially the long and complicated production pathway for the polysaccharide.

Transgenic Fish

By using different transgenic techniques, researchers are seeking to improve the genetic traits of the fish used in aquaculture. Researchers are trying to develop fish which are: larger and grow faster, more efficient in converting their feed into muscle, resistant to disease, tolerant of low oxygen levels in the water, and tolerant to freezing temperatures. For example, some species of fish make a protein which allows them to survive in the Arctic. This "antifreeze" gene has been transplanted into other species of fish so they also can survive in very cold waters. Growing fish that are longer and heavier is the goal of researchers who are experimenting with applying various types of growth hormone to fish. One method of doing this is to dip the fish in a solution which contains the hormone.

However, there are some problems with this technique. First, it may be difficult to produce large quantities of purified growth hormone, the method is labor intensive, and it's difficult to determine if the fish are getting the right amount of growth hormone. Therefore, researchers want to develop new strains of transgenic fish which naturally produce just the right amount of growth hormone to speed their growth. Such fish would be more cost-effective since they would produce higher levels of growth hormone on their own, and they would pass this trait to their offspring. There has been some success in this area. For example, a researcher at the University of Connecticut has developed a tilapia fish that grows twice as fast and up to five times as large as wild strains.

The scientist introduced an extra copy of the growth hormone gene into fish embryos at a very early stage, resulting in the unique growth characteristics. There are two main techniques which researchers use to transfer genetic material in fish. One is called microinjection, in which the genetic material is injected into newly fertilized fish eggs. However, this method is timeconsuming, so researchers may prefer to use electroporation. This involves transferring the genetic material, or DNA, into fish embryos through the use of an electrical current.

Better Fish Feed

Biotechnology is also helping to answer some of the technical and environmental concerns of fish farming. Many of these centre around what the fish eat. Right now the most common protein source for many fish diets is fish meal. Fish meal, a byproduct of fish processing, is used because of its high quality and high protein content. However, it has some disadvantages. One disadvantage for fish producers is that it is expensive. It sells for about $800 to $1,200 a tonne. So any cheaper alternative protein source would be welcomed. Another concern regarding fish meal is the stability of supply. Fish meal comes from the byproducts of wild fish, but world fish stocks are declining.

At the same time, fish farming is on the rise, and demand for fish meal is increasing. Given these factors, it's unlikely there will be enough wild fish to meet the increasing demand for fish meal. The use of fish meal in aquaculture causes other environmental concerns, as well. It contains levels of phosphorus far beyond the requirement for optimal growth in fish. The excess phosphorus goes into the water, causing problems such as eutrophication or excess algae growth. As a result of these concerns with fish meal, researchers are using biotechnology to produce alternative plant-based protein sources.

Plant protein has the potential to address the problem of phosphorus pollution, since plants do not contain such high phosphorus levels. As well, the use of plant protein in aquaculture would help take the pressure off wild fish stocks. Research in this province is focusing on the investigation of local crops as new sources for fish feed protein.

Some of the potential fish meal replacements include distillers byproducts, pulse crops and canola. Wheat, canola and canola oil are already being used to some extent in feed for aquaculture. Fish are very efficient in their growth, requiring much less energy than other animals. In the lab, researchers are able to give fingerling fish 700 g of feed and obtain 1,000 g of growth. That's because most of the growth is muscle, which is mainly water.

Because fish have such low energy needs, they can use high levels of dietary protein, often up to 40 per cent to 50 per cent of the diet. For Prairie crops to be used as the main protein source for fish, they must be processed into a concentrate. Biotechnology is often used in this processing. Plant protein also requires processing because plants contain what are called anti-nutritional compounds as a defence mechanism. These compounds must be destroyed during processing, or they could harm the fish. Researchers are also trying to deal with these antinutritional factors by producing feed enzymes to counteract them. Phytase is one example. This enzyme would help fish make the best use of the phosphorous available in a plant-protein based feed.

Pharmaceuticals

Therapeutic natural products found in terrestrial plants and microorganisms were the basis of early drug development and continue to be explored and used as sources of new drugs, albeit with declining rates of success. By contrast, exploration and use of marine resources, especially microorganisms (bacteria, fungi, and microalgae), for drug development has barely begun. Marine natural products are likely to provide a major new chapter in the history of drugs. Many bioactive substances from the marine environment already have been isolated and characterized, several with great promise for the treatment of human diseases. The compound manoalide from a Pacific sponge, for example, has spawned more than 300 chemical analogs, with a significant number of these going on to clinical trials as anti-inflammatory agents.

To promote sustainable technology development, governments should support research leading to new methods for discovery. One approach would be to focus on learning about natural functions, regulation, and production of substances generated by marine organisms, in order to identify potentially important agents. Refined test systems should be developed in order to identify selective agents produced by marine organisms.

Rapidly developing assay technology can facilitate exploration of the bioactivity of newly discovered compounds. These assay methods, which employ specific receptors for known physiological agents, require only minute amounts of a test substance and can be automated. Traditional chemical tests require considerable amounts of test material and laborious measurement processes. The new tools will make it feasible to test newly discovered compounds rapidly by the hundreds, for a wide spectrum of biological activities.

To date, exploitation of natural agents from the sea has been hindered by problems with limited or sporadic distribution and production. Much more research must be conducted to determine what seasonal factors and life cycle or reproduction states are linked with natural production of an agent. Factors influencing production may include diet, physical and chemical conditions, distribution by phylogenetic affiliation, geographic location, water depth, or associations with symbiotic microorganisms. Knowledge of these factors will be important in developing methods for producing selected metabolites, either from whole organisms or in vitro from cell or tissue cultures of plants and animals. Many of these compounds are very large and complex molecules, requiring very elaborate biochemical processes; as a result, it can be difficult to synthesize them or clone all the genes for standard production through fermentation.

Enzymes

Enzymes produced by marine bacteria are important in biotechnology due to their range of unusual properties. Some are salt-resistant, a characteristic that is often advantageous in industrial processes. The extracellular proteases are of particular importance and can be used in detergents and industrial cleaning applications, such as in cleaning reverse-osmosis membranes.

Vibrio species have been found to produce a variety of extracellular proteases. Vibrio alginolyticus produces six proteases, including an unusual detergent-resistant, alkaline serine exoprotease. This marine bacterium also produces collagenase, an enzyme with a variety of industrial and commercial applications, including the dispersion of cells in tissue culture studies.

Other research has demonstrated the presence in algae of unique haloperoxidases (enzymes catalyzing the incorporation of halogen into metabolites). These enzymes could become valuable products, because halogenation is an important process in the chemical industry. Japanese researchers have developed methods to induce a marine alga to produce large amounts of the enzyme superoxide dismutase, which is used in enormous quantities for a range of medical, cosmetic, and food applications.

An unusual group of marine microorganisms from which enzymes have been isolated are the hyper-thermophilic archaea previously called archaebacteria, which can grow at temperatures over 100° C and therefore require enzyme systems that are stable at high temperatures. Archaea typically are found in extreme environments, such as hot springs, animal guts, hydrothermal vents, sewage sludge digesters, and hypersaline habitats, including the Great Salt Lake.

Thermostable enzymes offer distinct advantages, many still to be discovered, in research and industrial processes. Thermostable DNA-modifying enzymes, such as polymerases, ligases, and restriction endonucleases, already have important research and industrial applications. Hot springs in Yellowstone National Park provided the first archaeon

(Thermus aquaticus) from which thermostable DNA polymerases were isolated. These novel enzymes (the Taq polymerases) became the basis for the polymerase chain reaction (PCR), a useful technique for studying genetic material. In 1989, thermostable DNA polymerase was designated Molecule of the Year by Science magazine. Comparable enzymes continue to be discovered.

Most enzymes involved in the primary metabolic pathways of thermophilic bacteria and archaea are dramatically more thermostable than are their counterparts living at moderate temperatures. Expanded study of enzymes from thermophilic marine microorganisms will contribute to the understanding of mechanisms of enzyme thermostability and should enable the identification of enzymes suitable for industrial applications as well as modification of enzymes to enhance thermostability.

Biomolecular Materials

Recent research has demonstrated that marine biochemical processes can be exploited to produce new biomaterials. For example, a corporation in Chicago is commercializing a new class of biodegradable polymers modelled on natural substances that form the organic matrices of mollusk shells. Equally exciting are the mechanisms used by marine diatoms, coccolithophorids, mollusks, and other marine invertebrates to generate elaborate mineralized structures on a nanometer scale (less than a billionth of a meter in size). Nanometer-scale structures can have unusual and useful properties. Research that will enhance understanding and allow engineering of the processes for creating these bioceramics promises to revolutionize the manufacturing of medical implants, automotive parts, electronic devices, protective coatings, and other novel products.

Biomaterials also hold promise for counteracting biofouling, which long has been recognized as an extensive and costly problem. Bacterial biofilms form slime layers that increase drag on moving ships, interfere with transfer on heat exchangers, block pipelines, and contribute to corrosion on metal surfaces. Bacterial and microalgal colonization of surfaces is accompanied by settlement of invertebrate larvae and algal spores, eventually leading to "hard fouling" and the need for costly cleaning.

The most effective anti-fouling coatings have utilized toxic chemicals, such as copper and organotins. There is an urgent need for non-toxic biofouling control strategies, due to heightened recognition of the impact that toxic coatings can have on the environment. Research is needed on the attachment mechanisms of marine organisms and the natural products they employ to prevent fouling of their own surfaces.

Molecular approaches to characterizing biofilm structure and development offer considerable potential for finding novel biofouling prevention strategies. It is now possible to determine the genes and pathways involved in regulation and synthesis of bacterial adhesive polymers. Considerable progress has been made in understanding the nature and expression of surface polymers produced by microorganisms such as the nitrogen-fixing Rhizobium species and the opportunistic pathogen Pseudomonas aeruginosa. Similar approaches can been applied to marine biofilm bacteria, to find the genetic determinants of adhesive production and the environmental factors that regulate synthesis.

Molecular biology techniques also can be applied to determine the basis of natural antifouling mechanisms. Many marine plants and animals remain free of attached bacteria, either because they produce repelling compounds or because their surface structure neutralizes bacterial adhesives. A product generated by the seagrass Zostra marina (eelgrass), for example, is an effective agent for preventing fouling by bacteria, algal spores, and a variety of hard-fouling barnacles and tube worms.

Molecular characterization of natural fouling resistance could provide new strategies for fouling control. Potential applications include prevention of fouling in industrial pipelines or heat exchangers, improved design of trickling filters or aquaculture circulation systems, and control of biofilm infections of medical implants and prosthetic devices.

Biomonitors

Marine organisms can provide the basis for development of biosensors, bio-indicators, and diagnostic devices for medicine, aquaculture, and environmental monitoring. One type of biosensor employs the enzymes responsible for bioluminescence.

The lux genes, which encode these enzymes, have been cloned from marine bacteria such as Vibrio fischeri and transferred successfully to a variety of plants and other bacteria. The lux genes typically are inserted into a gene sequence, or operon, that is functional only when stimulated by a defined environmental feature. The enzymes responsible for toluene degradation, for example, are synthesized only in the presence of toluene.

When lux genes are inserted into a toluene operon, the engineered bacterium glows yellow-green in the presence of toluene. This genetically engineered system "reports" that biodegradation of a specific chemical, in this case toluene, is proceeding.

Another type of biomonitor that holds great promise is the gene probe, which can be used to identify organisms that pose health hazards or may be useful in research. Specific gene probes can be employed, for example, to detect human pathogens in seafood and recreational waters; fish pathogens in aquaculture systems; microorganisms capable of mediating desired chemical transformations (e.g., toxic chemical degradation, CO_2 assimilation, metal reduction); and specific fish stocks in fish migration and recruitment studies.

Biopesticides

Natural marine products have the potential to replace chemical pesticides and other agents used to maximize crop yields and growth. Continued government support for R&D in this area is likely to result in useful natural pesticides that would provide greater specificity and fewer harmful side effects than do conventional synthetic agents. Current U.S. expenditures for all pesticides amount to $47 billion annually; by the year 2020, biopesticides from marine and other sources are expected to capture an estimated 10 percent of this market.

An example of a marine biopesticide in use today is PadanTM, which was developed from a bait worm's toxin known to ancient Japanese fishermen. This natural pesticide has demonstrated activity against larvae of the rice stem borer, the rice plant skipper, and the citrus leaf miner, among other pests.

More recently, scientists in Montana discovered novel compounds in marine algae and marine sponges containing symbiotic microorganisms. These compounds promoted growth and stimulated germination and increased root and coleoptile lengths in test plants.

Several sponge and nudibranch species produce terpenes, a broad class of aromatic compounds used in solvents and perfumes and known to deter feeding by fish. Extracts derived from these same sponge and nudibranch species also demonstrated powerful insecticidal activity against two species, grasshoppers and the tobacco hornworm.

Biomass for Energy Production

Approximately 40 percent of all primary energy production, or photosynthesis, occurs in the seas. In this process, oceanic plants (phytoplankton, seaweeds, seagrasses) take up carbon dioxide (CO_2) and, with light energy from the sun, convert it into organic carbon (primarily sugars) and oxygen.

The oceans contain 50 times as much carbon dioxide as does the atmosphere, and it is estimated that primary production incorporates 35 gigatons (1 gigaton = 1 x 1015 grams) of carbon into marine biomass annually. This abundant source of fuel for energy production has not been tapped commercially because it is not competitive with soybean meal and other easily harvested, traditional sources of biomass, and also because, regardless of the source, biomass is not competitive with other types of fuels. Governments should continue to support research on the use of biotechnology to enhance biomass production and utility. At least three general approaches are being explored.

First, the enzyme that captures CO_2 for photosynthesis— ribulose bisphosphate carboxylase\ oxygenase or "RUBISCO" — is relatively inefficient, so supercomputers are being used to verify structural information, and the enzyme is being redesigned to optimize its function.

Second, the chemical composition of biomass can be altered to make it more suitable for particular applications. For example, marine microalgae are being genetically engineered to boost their lipid content, with the aim of providing a source of alternative fuels that is more economical than are conventional sources.

Third, biotechnology is being used to convert biomass to ethanol and other alternative forms of energy and chemical feedstocks.

Improved Processes from the Seas

Bioremediation

Bioremediation shows great promise for addressing problems in marine environments and in aquaculture. These problems include catastrophic spills of oil in harbours and shipping lanes and around oil platforms; movement of toxic chemicals from land, through estuaries, into the coastal oceans; disposal of sewage sludge, bilge waste, and chemical process wastes;

reclamation of minerals, such as manganese; and management of aquaculture and seafood processing waste.

The full potential for marine organisms and processes to contribute new waste treatment and site remediation technologies cannot be realized without enhanced understanding of the unique conditions in marine environments. For example, oxidation- reduction (redox) states can fluctuate in coastal and estuarine sediments. The impact of changing redox conditions on biodegradation of environmental contaminants must be understood before waste management and remediation strategies and predictive models can be developed for contaminated sediments.

Bioprocessing

The emerging discipline of bioprocess engineering involves the application of biological science in manufacturing, to produce products such as biopharmaceuticals and natural bioactive agents. Bioprocess engineering requires an understanding of the biological system employed (such as a marine organism), isolation and purification of a product, and translation of the product into a stable, efficacious, and convenient form.

An emerging area of interest is the potential of marine bacteria and fungi to produce unusual chemical structures with no parallels in terrestrial organisms. Small-scale studies have begun to indicate the richness of marine microorganisms as sources for novel lead structures. The study of such materials is impeded, however, by the corrosive nature of seawater in conventional fermentors. The use of marine organisms as a basis for bioprocess engineering depends on the development of salt-resistant reactors. In addition, numerous technical advances are required for progress in bioprocess engineering in general. For example, improved methods are needed for the isolation and purification of products from the producing organisms. Advances in this area most likely will come from emerging partnerships between the government and the private sector.

Enhancing Reproduction and Early Development

Biotechnology can be applied to enhance reproduction and early development of cultivated aquatic organisms. The resulting benefits could include year-round production of gametes and fry of economically valuable species and creation of new markets for specialized, genetically improved broodstock. Similarly, biotechnology may provide techniques for improving the reproductive success and survival of endangered species, thereby helping to preserve the diversity of life on Earth.

As a first step, research should be directed toward improving basic understanding of environmental, hormonal, biochemical, and genetic control of reproduction. More specifically, scientists must identify and understand the mechanisms of expression of genes involved in reproduction and development, improve technologies for cryo-preserving gametes and embryos, improve delivery systems for administration of natural and synthetic hormones, and enhance understanding of the pharmacokinetics of uptake and release of administered hormones.

Improving Growth, Development, and Nutrition

Biotechnology can be applied to improve the growth, development, and nutrition of cultivated aquatic organisms. Specifically, the tools of biotechnology can enhance understanding of the environmental, biochemical, hormonal, and genetic bases of growth and nutrition for all life stages. Biotechnology also can be used to expand knowledge about the absorption, digestion, and assimilation of nutrients and to increase the efficiency of these processes.Governments should support research in two key areas. First, recombinant DNA technology should be exploited to produce transgenic aquatic organisms with enhanced growth and food conversion characteristics. In addition, biotechnology should be applied to improve methods for administering hormones, nutrients, and other growth-enhancing substances while enhancing understanding of the pharmacokinetics of uptake and residues of administered substances.

Improving Health and Well-being

Biotechnology offers substantial opportunities to improve the health and well-being of cultivated aquatic organisms. More than 50 diseases affect fish and shellfish cultured in the United States, causing losses of tens of millions of dollars annually. Biotechnology not only can improve the survival, growth, vigor, and well-being of cultivated stocks, but also can reduce disease transfer between cultivated and wild stocks. New products and market opportunities can be developed related to aquatic animal health and well-being.

The tools of molecular biology can provide a basic understanding of host immunity, resistance, and susceptibility to diseases and associated pathogens by furnishing information about life cycles and mechanisms of pathogenesis, antibiotic resistance, and disease transmission. Improved technologies must be developed for detecting and diagnosing pathogens and diseases and for enhancing the genetic basis of disease resistance, thereby reducing the need for antibiotics and other drugs.

Potential products resulting from this research include gene therapy techniques; broodstock free of pathogens; safe, effective prophylactic agents, including immune modulators, antigens, and vaccines; safe, effective therapeutic agents; and improved systems for administering prophylactic and therapeutic agents.

Improving Quality and Value

Governments have the responsibility to help ensure the safety and quality of food supplies, and biotechnology can and should be an invaluable tool in carrying out this mandate. Biotechnology can be employed to assess and improve the safety, freshness, color, flavor, texture, taste, nutritional characteristics, and shelf life of aquacultural food products. In addition, practical technologies can be developed to detect and assay toxins, contaminants, and residues in seafood, and to reduce or eliminate contaminants. There are also opportunities to apply biotechnology in improving seafood processing. Research should be conducted to develop and improve technologies for all these applications.

Conserving Genetic Resources

The preservation and enhancement of biodiversity in natural systems is an important priority. Therefore, governments should encourage and support programs to maintain and enhance biodiversity in aquatic systems through cultivation and stocking of aquatic species. Biotechnology can be employed in two ways to conserve genetic resources of aquatic species.

First, the tools of biotechnology can be used to identify and characterize important aquatic germplasm, including endangered species. Genomes of aquatic species can be analyzed and characterized, and quantitative trait loci identified. Second, biotechnology can be applied to improve understanding of the molecular basis of gene regulation and expression as well as sex determination and thereby improve methods for defining species, stocks, and populations.

Approaches include developing marker-assisted selection technologies, improving precision and efficiency of transgenic techniques, and improving technologies for the cryo-preservation of gametes and embryos. Ultimately, stocking certain areas with selected, cultivated species and strains could help maintain biodiversity in natural aquatic ecosystems.

Enhancing Biomedical Models

Aquaculture has important purposes other than food production. Because they often adapt to extreme environments, marine organisms can provide unique models for research on biological and physiological processes. Studies of the developmental, cellular, and molecular aspects of marine organisms as model systems will provide insights into the basis of disease mechanisms and pathogenesis in humans. By contrast, use of mammalian organisms as a basis for the development of some types of human disease models may be neither feasible nor cost effective.

Progress in this arena will require that sophisticated molecular biology technologies be adapted to marine organisms, in order to enhance understanding of their biological processes. For example, approaches for gene transfer into eggs have been developed for many terrestrial organisms, but not for most marine species. This technology is needed for analyses of gene regulatory systems and gene expression.

In addition, methods need to be developed for culturing tissues from marine organisms. Cultured cell lines will provide opportunities for gene transfer and gene expression studies and enhance the usefulness of marine species as biomedical research models. This is an important research area deserving of government support.

Conserving the Seas Using Biotechnology

Scientists have a powerful new array of sampling devices and measuring instruments that will accelerate greatly the acquisition of knowledge about ocean resources and foster their wise use. These technologies include manned deep-sea submersibles, remotely operated vehicles, geosynchronous satellites, sophisticated acoustic measuring devices, pressure—retaining deep-sea samplers, geographic information systems, real-time flow cytometry, PCR and biomonitoring techniques, computerized databases, and other forms of information exchange and analysis.

These tools should be exploited to accelerate the discovery of unknown marine microorganisms and to expand understanding of known varieties. Government support for this research is essential, because only then will sufficient information be acquired to assure that practical applications will result. As new life forms and processes become known, and as understanding of them grows, marine biotechnology will make significant contributions to the nation's social and economic well-being.

Identifying Organisms and their Niches Biodiversity

Nowhere in the biosphere is biological diversity greater than in the seas, and the extent of this diversity becomes increasingly evident as scientists investigate new environments. In the 1980s, for example, giant tube worms (Riftia) were recovered from areas adjacent to deep-ocean thermal vents, and novel mussels (Bathymodiolus) that farm methanotrophic bacteria on their gill tissue were discovered around methane seeps in the Gulf of Mexico. Most newly described species have been and will continue to be microorganisms, although it is clear that new marine plants and animals also await discovery.

Less than 1 percent of the extant bacteria — marine and terrestrial — have been isolated and described. The marine environment represents a particularly fertile source for new bacteria, as evidenced by the recent discovery of unusual "cold water" archaea 100-500 meters deep in the oceans. These archaea comprise a high percentage of the total bacterial ribosomal RNA present in seawater samples, yet they have not been isolated in pure culture and described. Pursuit of this research may provide new understanding of oceanic processes and, once these archaea are cultivated, perhaps a novel source of products. Countless other marine bacteria also have yet to be cultured; advances in microbial culture technology will enable scientists to isolate, describe, and possibly exploit these organisms.

Viruses are another newly appreciated element of marine biological diversity. Electron microscopy and PCR technology have revealed abundant numbers of viral particles in seawater samples. Specific viruses that infect species of marine phytoplankton have been cultured from coastal as well as open ocean sites. This discovery is exciting for at least two reasons. Viral infection of higher forms of marine life almost certainly affects global ocean processes, such as photosynthesis. Marine viruses also will provide new materials for development of genetic and biotechnological tools that can be used to study and manipulate marine organisms. For example, marine viruses could be used to genetically engineer higher forms of marine life.

Marine Ecology

As a basis for developing new applications for marine products and processes, analyzing global climate change, and improving fisheries management, scientists must build a fundamental understanding of marine organisms and their specific adaptations to and interactions with their environment. Recent developments in molecular techniques are rapidly expanding capabilities for research on marine ecology. Three research areas seem to be especially fertile and deserving of government support.

As a first step, the new molecular tools should be applied to gain insight into the basic molecular and cellular processes by which marine organisms adapt to extreme environments. Examples of basic research that may lead to commercial applications are gene sequencing projects and the ongoing study of special glycoproteins that inhibit ice crystal growth in the tissues of Antarctic fish. These substances, along with gene sequence information for key marine species, may prove useful in industrial and medical preservation processes.

Second, a thorough understanding of marine ecological systems must be developed in order to specify the "normal" baseline level of function and to monitor and predict potential changes and perturbations of systems due to physical, chemical, or biological impacts. The development of predictive models for analyzing potential global climate changes depends on the acquisition of fundamental information on molecular regulatory mechanisms of photosynthesis in the oceans.

Third, research on marine ecology can be conducted to benefit fisheries management. The tools of biotechnology can be used to determine the effect of natural and anthropogenic perturbations on the size of commercial stocks, to delineate fishery stocks and management units of living resources, to determine predator-prey relationships, and to restore habitats essential to robust fisheries.

Powerful tools are being developed to elucidate the many biogeochemical cycles that determine the fate of all the life- supporting elements on Earth. Scientists are beginning to understand and manipulate the molecular genetics and biology of esoteric metabolic pathways associated with the carbon, sulfur, phosphorous, iron, and other biogeochemical cycles. For example, based on the hypothesis that iron controls photosynthesis in the oceans, immunological probes were used to show that addition of iron to open ocean water off the Galapagos Islands significantly increased energy production.

Marine biotechnology will be useful in assessing the role of the oceans in affecting climate change and the global carbon cycle. Molecular techniques can facilitate and enhance the measurements of CO_2 concentrations and total CO_2 inventories being developed for global ocean models of carbon cycling. There is compelling evidence that the exchange of dissolved and particulate materials between the continental shelf and its boundaries is a significant factor affecting the flux of CO_2 and biogenic elements within the global ocean. Several international agencies plan to collaborate on related research, including marine biotechnology applications.

References

Houdebine, L.M. 2003. *Animal transgenesis and cloning*. New York, John Wiley and Sons.

National Research Council (NRC). 2002. *Animal biotechnology: science-based concerns*. Washington, DC, National Academies Press.

Pew Initiative on Food and Biotechnology. 2003. *Future fish: issues in science and regulation of transgenic fish*. Washington, DC, Pew Initiative on Food and Biotechnology. 72 pp.

Sagar, A., Daemmrich, A. & Ashiya, M. 2000. The tragedy of the commoners: biotechnology and its publics. *Nature Biotechnology*, 18 (January 2000): 2-4.

7

Conservation of the Genetic Resources of Fish

The need for conservation of fish genetic resources has been recognized by fishery scientists and aquaculturists for some time, especially in relation to overfishing of natural stock, the effects of large-scale alterations to river systems, and domestication of species through aquaculture. The FAO World Symposium on Warm Water Pond Fish Culture stressed the importance of genetic selection and hybridization in improving fish varieties for culture and noted problems of excessive inbreeding on carp farms. Need for an international system of designating strains and stocks was also recognized. In 1971, FAO established an ad hoc Working Party on Genetic Resources of Fish, which reviewed progress in genetic selection for fish farming, identified priority areas for research and made several recommendations for the conservation of fish genetic resources. Included were suggestions that methods for doing so needed urgent study, that a catalogue of threatened genetic resources of potential use to aquaculture should be prepared, and that the collection of wild species of potential usefulness should be undertaken. These actions have not taken place, largely because there has been no consensus on the criteria that are needed to limit such work to manageable proportions. The FAO Technical Conference on Aquaculture, held in Kyoto, Japan, 26 May to 2 June 1976, reaffirmed the need to maintain the genetic diversity in artificially propagated stocks, noted that indiscriminate transfer of fish and shellfish have, in some cases, had adverse effects on indigenous stocks, and called for increased research on fish genetics as there was a substantial lack of information on this subject, making the formulation of selective breeding programmes difficult.

One strategy for conserving the diversity of natural populations of fish and other organisms is to set aside aquatic reserves. The International Union for Conservation of Nature and Natural Resources (IUCN) has been active in promoting marine parks, and with the cooperation of the Government of Japan, the World Wildlife Fund and UNEP, held an International Conference on Marine Parks and Reserves in Tokyo, 12 to 14 May 1975. This conference reviewed the conclusions and recommendations of earlier regional and world conferences concerned with the marine environment and proposed, inter alia, that IUCN and other concerned agencies "develop a coordination strategy to satisfy forthcoming requests for assistance [to developing countries in the establishment of marine parks]", and "set up teams to undertake appropriate surveys to enable systems of marine parks and reserves to be based on the best available data, to assist in the formulation of appropriate conservation policies relating to marine parks and reserves, and to identify those projects suitable for bilateral and/or other technical cooperation programmes".

The United Nations Conference on the Human Environment, Stockholm, 1972, produced recommendations emphasizing, inter alia, the need for conservation of all genetic resources. In its third session the Governing Council of the United Nations Environment Programme requested, among other things, the preparation of an overview on the problems of the conservation of genetic resources as part of an overall report on the conservation of genetic resources. The Food and Agriculture Organization, through a cooperative project with UNEP (FP/1108-75-01) prepared a paper which included a brief review of the problems and requirements for action on fish resources (UNEP/PROG./4). This overview has now been revised and produced by UNEP as UNEP Report No. 5 , "An Overview of Genetic Resources". The chapter emphasized that the basic constraint is lack of knowledge and recommended, among other things, the need for a mechanism for monitoring changes in the genetic diversity of fish populations, for promotion of research directed at creation of knowledge on the genetics of fish which would assist in a more applicable definition of genetic impoverishment in fish species and for promotion of research on appropriate methodologies for conservation.

During discussions held in FAO, Rome, in June 1979, it was agreed that the variety of needs and problems associated with this aspect of genetic conservation could not be adequately reviewed and assessed by a single individual. It was proposed that a group of experts be convened to accomplish this review and propose a balanced and practical programme of action.

It was then agreed that FAO would organize an expert consultation, in cooperation with UNEP, and with the participation of Unesco and IUCN, to be held in Rome, 9–13 June 1980.

The purpose of the consultation, as stated in a prospectus sent to prospective participants, was:

1. to consider in depth the problems necessitating the conservation of fish genetic resources;
2. to develop a scientific consensus on the needs and on a strategy to solve these problems;
3. to propose an action plan and appropriate methodology to carry it out.

Subsequently, it became evident that the consultation needed to focus primarily on the scientific aspects of the problem, considering especially the nature of the threats to genetic diversity of fishes, both among and within species, methodologies for monitoring and assessing change in genetic diversity, and identifying feasible actions which might be taken to conserve diversity. The development of a comprehensive and coherent action plan for implementation by governments was seen as a subsequent step. Thus, in the course of the consultation, the experts were urged to address recommendations to several different kinds of people or institutions: scientists, conservationists, fishery managers and aquaculturists, governments and international organizations.

Nature Conservation Versus Genetic Resource Preservation

The objective of nature conservation is to preserve for posterity as much as possible of the earth's biological and ecological diversity. The justifications of such an attitude span a very broad range of human thought and experience - from divinely inspired ethics to short-term economic and political gain. At the one extreme some people argue that morality dictates the right of all species to persist and evolve, and that man's power to alter and destroy the biosphere does not give him the ethical licence to do so. On the other hand, there are those who argue from the standpoint of human welfare - that the preservation of biological diversity is only common sense, and that the prudent strategy is to keep the options open by preserving genetic diversity in order to sustain production in agriculture, forestry, fisheries and other natural products and to minimize the threats of ecological disasters such as climatic alteration, erosion, siltation, desertification and pollution.

While there is no "best" approach or justification, there are appropriate ones. Genetic resource preservation is one such tactic. The approach taken by proponents of Genetic Resource Preservation is pragmatic. As a matter of human welfare, it is essential that production of natural resources increases. Fish are an important source of protein and other valuable organic products. Therefore, the protection and exploitation of options for the improvement of fisheries and aquaculture is a high social priority. The accomplishment of these goals depends to a large and increasing extent on technology and science and it is the role of genetics in this effort to which this chapter is directed. An interest in and knowledge of the role of genetics in the enhancement of fisheries production does not necessarily make one a conservationist. But such knowledge will make a more effective soldier in the battle to feed the rapidly growing human population.

Extinction Versus Genetic Impoverishment

It is convenient and heuristic to distinguish two processes by which genetic resources are lost; these are (1) the extinction of a species and (2) the reduction of genetic variation within a species. The former process, once it occurs, is qualitative, final and irreversible. The latter process is a matter of degree, and is to some extent reversible.

This chapter discusses aspects of both processes because both are clearly relevant to the maintenance of options in fisheries. Nevertheless, the relevance to exploitation of these two processes depends on the stage of development, its intensity and the degree of human control in the fish extraction enterprise. These problems will be dealt with throughout the chapter.

A realistic perspective on the process of extinction and its current status is essential for those in the resource fields. Of special import is the current change in rate of extinction. Barring periods of dramatic climatic swings such as the Pleistocene, both the origin and extinction of species have probably gone on at similar rates for hundreds of millions of years. Now, however, due to habitat destruction by man, extinction rates are probably three or more orders of magnitude higher than ever before, while the rate at which new species of large organisms appear, especially in the tropics, is approaching zero . Hence for groups of organisms that occupy disappearing habitats, it is not likely that organic diversity will ever again reach its present levels.

For fishes the situation may not appear to be so severe, but one must specify the place and the group. Habitat destruction in the oceans is not yet appreciable. At least no significant or detectable increase in extinction rates has been observed in the oceans (although populations have been extinguished from overfishing and pollution). For other aquatic habitats, however, the situation is already deteriorating rapidly. The issue boils down to one of disparate time scales. The time scale for pollution, exploitation or development is measured in years, decades or centuries, whereas that for speciation is usually measured in much longer intervals.

For example it may only require ten years to dam a river, thus creating a habitat inimical to the existing riverine fauna dependent on periodic flooding or moving water. On the other hand, the lifetime of the resulting reservoir will be too short by thousands of years for the evolution of lacustrine forms of fish.

Similar arguments could be made for the effects of pollution and siltation on rivers, lakes, estuaries, lagoons and reef habitats. Obviously, it is naive to entertain the sanguine expectation that evolution of new species will compensate for the loss of biological variety during the next few decades.

The Case for Genetic Resource Preservation in Fish

The assumption that underlies the discussion in this section is that the manner in which a resource is exploited is as important in the long run as the degree to which it is exploited. More explicitly, the way in which fish populations are managed will largely determine the success of the enterprise. This comes as no surprise to fisheries biologists, but there is a dimension to management that has been relatively ignored - the genetic dimension. One of the objectives of this chapter is to underscore this dimension and by so doing, to anticipate and thereby prevent the very serious loss of options that has already occurred in agriculture, and to a lesser extent in animal husbandry.

Nutritional arguments for Genetic Resource Preservation

Fish flesh is an important source of high grade animal protein. With a growing world population and rising expectations for living standards, the pressures on fish both as a source of food and for its byproducts can only increase in the future. At present fish and aquatic animals constitute 17 percent of the total animal protein in the human diet. This statistic conceals wide regional differences in fish consumption. For example, 32 countries obtain 34 percent or more of their animal protein from seafood. On the African continent, ten countries obtain over 40 percent of their protein from fish. In 21 countries on the same continent over half of the fish landed comes from inland lakes and rivers, and in 13, the fish supply is entirely from inland sources.

The potential for further growth of natural marine and freshwater fisheries is limited. Already the total world harvest is 15–20 million tons lower than it might have been had fishing been strictly controlled through scientific management, and 25 valuable fisheries have been seriously depleted by overfishing. Other factors which limit the contribution of marine fish to the diet include problems of distribution, storage and cultural inhibitions to its consumption.

Intensive fish farming can, in principle, overcome two of these limitations. Fish farming activities may often be sited near to the markets in which the product is sold. They also avoid the depletion of biological resources except where still dependent on the collection of seed from the wild. Fish farming not only allows the products to be delivered directly to their markets but it also represents both a renewable and inexhaustable resource. Aquaculture is still in a relatively early stage of development and the production of new strains and races of food fishes will require careful management of available genetic resources. The lesson learnt from extensive breeding in crop plants and in domestic animals is that a narrowing in the genetic base of the species is inevitable. In such processes, genetic determinants which are likely to be lost at an early stage include those controlling disease-resistance and fitness in marginal environments. It is hence prudent for breeders to be especially mindful of the need to protect and preserve, at the earliest stage possible, broad genetic diversity within those species most likely to come under intensive breeding pressure.

Economic arguments for Genetic Resource Preservation

Fisheries not only provide a significant portion of the protein available for human consumption -they are also an economically significant activity, providing jobs and investment opportunities, and, for many countries, a means of improving the balance of international trade.

Certain species of fish are, by virtue of special or unusual biological features, useful as experimental animals and as a source of biochemical and pharmacological agents. Often species and genetic traits to be preserved are not the same as those for food fishes. Since only a handful of the 25 000 plus species of fish have been subject to scientific scrutiny the extinction of any species may represent a potential loss of an economic resource.

Substances isolated from fish and other aquatic animals are already widely used in medical research. For example, tetradotoxin (TTX), a toxin isolated from the puffer fish, Tetraodon immaculata, is used in neurophysiological research as a specific sodium channel blocker and is playing an important part in reaching an understanding of the basic ionic mechanisms of nervous transmission.

Another class of compound used as a research tool is luminescent proteins. For example, Aqueorin from a jellyfish species has been widely used to monitor the Ca^{2+} concentration in cells. Ca^{2+} serves as an important signal in transmitter/hormone release, and in excitation contraction and excitation-secretion coupling. An understanding of such processes is important in the development of new drugs and in the treatment of disease. It is likely that fish toxins, hormones, glycoproteins and naturally occurring polypeptides constitute a large and relatively unexplored reservoir of compounds of equal pharmacological interest.

Other natural products, notably oils and waxes, have long established industrial uses, e.g., for vitamin enhancement of animal feedstuffs. New applications for these byproducts might include substitute natural bases for medications and cosmetics. In many countries fish meal is already widely used for feeding domestic animals and as a major source of fertilizer.

In cases where species are threatened with extinction, establishing the importance of natural products remains an issue of high priority. The economic and social benefits provided by such discoveries, however, go far beyond the immediate value of the fish as an economic resource. The history of natural resource utilization tells us that the ultimate value of any genetic resource may not be appreciated at the present time.

Nor is it sufficient to stave off extinction by preserving a small handful of individuals. It is not uncommon that the genetic determinants are not evenly distributed within the species - in quantitative or even qualitative terms. This presents a strong argument for preserving broad genetic diversity within a species, if its socio-economic potential is to be realized.

Another important area of economic activity relates to sport fishing. Increasingly, freshwater sport fisheries are 'scientifically' managed and restocked from hatchery populations. Freshwater fishing in countries such as Ireland and marine sport fisheries in various tropical countries are a major factor in the development of tourism. Thus sport fisheries can increase employment and support and stimulate a wide range of ancillary economic activity. The creation of classes of marine reserves where controlled fishing is permitted may provide a valuable way of safe-guarding the species diversity of coastal areas while providing a good economic return from anglers and day visitors.

In North America, Europe and South-East Asia, the ornamental aquarium trade is a major industry. For example, the wholesale turnover of tropical fish in Florida (U.S.A.) alone in 1974 was in excess of $30 million . Many of the most popular species come from the Amazon basin and the Great Lakes of Africa. These fish often occur in small discrete sub-populations which live in highly specialized or localized habitats and are thus extremely vulnerable to extinction. Properly guided and encouraged, the culture of such species by hobbyists could be one way of preserving this resource for future generations. However, owing to the danger of inbreeding, in situ conservation is much more likely to be successful. Salt water coral reef fishes provide similar potential for income as well as similar problems for the long-term protection of the resource. Certain species of fish are now being seeded into lakes and rivers for use in the control of weeds and insect pests. This may have significance in the elimination of vector-borne disease. While this is socially and economically desirable attention should be paid to the effects on indigenous fish populations.

In summary, in order to keep our economic options open, it is desirable to implement aggressive programmes that operate on both levels of genetic resource preservation - programmes that (1) minimize the rate of species extinction by mitigating habitat disruption and that (2) minimize the erosion of genetic variability of species in no immediate danger of extinction, especially of those which already play major economic roles in human survival.

Ecological arguments for Genetic Resource Preservation

The principle ecological argument for the preservation of fish genetic resources is, in a sense, the logical converse of the preceding arguments. It is as follows: Stability of ecological systems and the maintenance of biological (taxonomic) diversity is a universally acknowledged value, although social and economic considerations may often be granted a higher priority. One important means of maintaining stability and diversity is the maintenance of fitness in species, particularly the dominant, high trophic level consumers. The depletion or extinction of such species constitutes the loss of genetic resources and is a threat to the integrity of ecosystems.

Habitat protection is the best way to insure the survival of ecologically important fish species. Habitat destruction is particularly disastrous when local endemic species are involved because the elimination of an entire species is an irreversible process. We expect this to be an increasing problem, particularly in tropical areas where endemics with narrow habitat requirements abound. While some decay in water quality and habitat diversity can be tolerated, there is probably a point at which even partial destruction of habitat integrity can reach genetic and ecological thresholds that result in the precipitation of extinctions. Some tropical ecologists believe that the extinctions of certain key species will bring about a rather sudden cascade of linked extinctions. In summary, a concern for survival of species, and the development of appropriate preservation programmes and guidelines, will indirectly benefit man by protecting entire ecosystems and all their attendant economic and ecological values.

Levels and Methods of Genetic Resource Preservation In Fish

From the point of view of the preservation of the genetic resources of fish, four main levels of concern and strategy can be identified: (1) oceanic systems, (2) continental waters (both fresh and marine), (3) aquaculture and (4) stock enhancement programmes. The last two levels of concern are directly controlled by man, whereas the first two can be regarded as uncontrolled systems, although human activities can have major impacts on them.

Human Impacts on Aquatic Habitats

There are a number of characteristics which clearly differentiate oceanic systems from continental waters as far as genetic resource preservation is concerned. Firstly, the much greater size of oceanic systems, in terms of both physical dimensions and biomasses involved, make them much more difficult to manage. It must be remarked, however, that a few of the larger river basins (e.g., in South America and Africa) and their exploited fish biomass approach oceanic dimensions.

The effects of meteorological and especially hydrological cycles are usually much more crucial in continental waters where they have a strong influence on fish populations - especially as far as reproductive migrations or strategies are concerned. The combination of these characteristics means that both natural and man-induced stresses have a relatively greater impact in continental systems. Included in this group are intertidal habitats, coral reef systems, brackish waters and fresh waters.

Biologically, these conditions give rise to two further distinctions which are relevant to genetic resource conservation: (a) continental areas have very much higher numbers of endemic and locally distributed species and sub-populations than oceanic systems, and (b) the lower impact of stresses on pelagic oceanic species, and their high fecundity, make their extinction unlikely. At low abundance levels their exploitation will be uneconomical, whilst in enclosed continental systems many species could, and indeed have, disappeared.

In oceanic systems, the main changes are produced by the direct but selective exploitation through fisheries. Such exploitation can have two different effects on a considered species or population. One is a direct effect on the population structure of the target species by selective removal of a particular sector. The other is an indirect effect on non-target species, where the fishery may affect the food chain by, for example, upsetting established interactive relationships.

In continental waters the same effects can occur but often to a much greater degree and whole populations can become extinct, solely through over-fishing. Here, however, other important stresses occur, notably pollution, disruption of life history cycles and disruption of ecosystem structure. Examples include the introduction of foreign predators, competitors and pathogens.

Among the different kinds of pollution, air-borne pollution leading to acid precipitation probably has had the most devastating (and sometimes selective) effect on fish populations particularly in southern Scandinavia and the north-eastern and north-central U.S.A. The direct pollution of waters through the discharge of various chemicals (some of them mutagenic) as wastes or the run-off of agricultural chemicals is also of major importance.

Land-use may also have highly undesirable effects such as the production of very acid soils (and run-off) and the silting of streams.

Another serious human impact is the disruption of biological (in most cases reproductive) cycles in fish populations as a consequence of building dams or other obstructions on rivers where migratory species occur. Dams may not only prevent the migration to upstream spawning grounds, but they may also change rivers into semi-lacustrine habitats quite unsuitable for stream species. Ecological barriers are also common and may be caused by zones of pollution in the lower reaches of rivers preventing the migration of various species.

The introduction of exotic species may also be included as a factor of importance to endemic fish populations. They may lead to the introduction of disease, higher levels of predation, or so affect the ecosystem (e.g., by competition for food) as to cause the extinction of local species. It is very rare to find introductions which have filled a totally empty niche, especially in the tropics, so introductions are highly likely to result in changes in the endemic populations.

Compared to oceanic and continental systems, aquacultural, hatchery and stock enhancement programmes are subject to an even greater degree of human control by definition. Within such systems, direct environmental and genetic management can and should be imposed and should only be limited by economic criteria and technological sophistication.

Ways and Means of Genetic Resource Preservation

A strategy for genetic resource preservation is largely dictated by the type of aquatic system. This section briefly outlines the approaches relevant to the various kinds of systems.

Oceanic systems

For any oceanic fish species, one of the first requirements for the formulation of a preservation policy must be a distribution map and the identification of any distinct sub-populations which there may be. Distribution data will probably be available from relevant fisheries, but the detection of sub-populations may have to be carried out by the use of one or more of the 'labelling' techniques available (e.g., tagging, the identification of meristic characters, electrophoresis, etc.). The importance of detecting sub-populations is that, although the danger of the extinction of oceanic species through overfishing is probably very small (the assumption being that the fishery will disappear before the species), it is perfectly possible that a unique sub-population could disappear. (Incidentally, the practice of a rushed, published description of a local population as a new sub-species, merely for the purpose of forestalling development, is a legalistic device that will probably prove counter-productive, since it is likely to be abused.)

Significant changes within an exploited population can only be detected through a continuing programme of monitoring designed to study the overall population structure, its composition and its change. The decline in any part of the population can probably only be controlled by regulation of the fishery, either through reduction of exploitation in space or time, or by altering the nature of the gear used.

The effect of fishing on the genetic diversity of oceanic populations is discussed below. If there is evidence of undesirable genetic changes in any species because of selection imposed by the fishery, it would be ideal to try to reverse the trend by appropriate management tools.

Continental waters

One of the enormous problems related to smaller enclosed systems, especially in the tropics, is the identification of the species involved. In parts of Africa, Asia, and especially in South America, large numbers of endemic and very locally distributed fish species occur and many are undescribed. Thus an essential pre-requisite to any broad programme of genetic resource preservation is a proper taxonomic study of the fish species occurring in each area and a full check-list of these species, indicating the status of each, and, if possible, its significance in ecological, economic, scientific and social terms. Such lists would provide the basis of any international list of endangered species to be included in the Red Data Book published by the IUCN.

Because of the high probability of the loss of genetic diversity in any species taken into culture, the best method of maintaining this diversity is to conserve self-maintaining populations in natural habitats. The normal procedure here would be to establish nature reserves (lakes, river basins, estuaries, coastal lagoons, coral reefs, etc.) which already contain one or more populations of the fish species or communities concerned. These reserves could well be multi-purpose areas, conserving several other types of habitat and communities in addition to the fish resource, e.g., National Park of Sabana Grande, Venezuela.

In some situations an individual species or its habitat may be so threatened that the only option is to collect stock and transfer this to an aquaculture system, or preferably, an alternative suitable location to create a new population. Ideally this new location should be within a nature reserve or in an area not liable to man-made pressures. The refuge so created can later be used to re-stock the original habitat if conditions there improve. This strategy is being used at present in both Canada and Scotland to preserve local populations of whitefish (Coregonus). Aquaculture techniques may be used on a short-term basis to enhance local stocks which are subject to temporary recruitment problems.

By definition, it is essential that man-induced stresses be strictly regulated in any established nature reserve. This includes control not only of fishing effort and gear, water quality degradation, barriers to migration, etc., but also of introduction of exotic fish or any other species likely to have a harmful impact on the ecosystem. The wild trout and charr watch concept, as recently proposed by Regier and Powers , may prove to be one useful way of developing international monitoring schemes for important species or groups of fish.

Once aquatic reserves are established, scientific management to prevent degradation and loss of diversity is absolutely essential. In terrestrial ecosystems, management of nature reserves is a new and often contentious discipline, but the high rate of extinctions for vertebrate species in terrestrial reserves requires the immediate attention of managers and consultants . Fishery and aquatic reserve managers would be well advised to consider this problem at the outset.

General Principles of Genetic Resource Preservation

The Importance of Genetic Variation

Genetic variation is the raw material in species populations which enables them to adapt to changes in their environment. New genetic variation arises in a population from either spontaneous mutation of a gene or by immigration from a population of genetically different individuals. Alternate forms of a particular gene (or locus) are called "alleles". The number and relative abundance of alleles in a population is a measure of genetic variation, sometimes termed "heterozygosity". Genetic variation is a measure of a population's ability to adapt to environmental change or stress, and thereby to survive.

Population geneticists have spent several decades establishing the importance of genetic variation in natural populations. It is known, for example, that the response to natural selection by experimental populations is accelerated by mutation-inducing radiation and/or the introduction of genes from different strains. In terms of the preservation of genetic resources, we can expect that benefits would be derived from maintaining the maximum level of genetic variation in a strain as well as by maintaining multiple strains that could serve as additional sources of genetic information by hybridization. It follows that the loss of genetic variation for whatever reason (e.g., prolonged selection, inbreeding, isolation) will result in a decrease of the potential adaptability of a population.

It appears that the benefits of multigene heterozygosity are universal in outbreeding organisms. In several organisms, including some fish species, individuals possessing the most genetic variation have been shown to have better survival rates or higher relative growth rates. Relatively heterozygous individuals appear to be more resistant to environmental perturbations during development. Clearly, genetically variable populations have many advantageous characteristics that are absent from genetically impoverished ones.

Over the past several years a rather large body of evidence has accumulated on the biochemical differences between alleles of genes coding for metabolically important enzymes. These biochemical differences emphasise the relationship between genetic and functional diversity. The functional properties of different alleles often reflect a biochemical and genetic adaptation to life in a heterogeneous environment. The extent, however, to which the states of physiological and behavioural traits at the whole animal level can be correlated with genetic and biochemical data is still uncertain. Most workers in the field agree that a proportion of protein polymorphisms have no direct or measurable effect on viability or some other aspect of fitness. Nevertheless, there is significant evidence from studies of individual genes, organisms and populations to substantiate the importance of genetic variation to population adaptability.

Effects of Inbreeding

The selection of small numbers of parents can reduce genetic variability. Equally serious is the fact that brood stock may be continually selected from closely related, perhaps full sib individuals. This leads to generation after generation of inbreeding of closely related individuals which very often results in homozygosity for unfavourable genes. The overall result is inbreeding depression.

Inbreeding depression is the loss of fitness (e.g. vigour, viability, fecundity) in connection with the loss of genetic variation due to homozygosity. The evidence that inbreeding is harmful is copious and virtually universal.

A general view of the effects of inbreeding and its relationship to the conservation of genetic resources can be found in Soule and is treated more extensively by Frankel and Soule . Quoting from the former: "A survey of inbreeding experiments leads to the generalization that increasing the inbreeding coefficient by 10 percent induces a 5–10 percent decline in a particular reproductive trait." Note that an F value equal to 10 percent approximates the amount of inbreeding that would theoretically occur in a population of five adults breeding at random for a single generation, or in a population of 25 adults breeding at random for five generations.

A 5–10 percent decline in fecundity might not appear to be very serious (especially when dealing with such fecund animals as fish) but if the effects of inbreeding depression on other traits (such as viability) are also considered this amount of inbreeding can lower reproductive potential as a whole by 25 percent (e.g., in fowl and swine). Gjederm showed that a 10 percent increase in the inbreeding coefficient in rainbow trout can result in a 10 percent decline in hatchability, and a 24 percent decline in viability of fingerlings. It should be noted that such independent effects are multiplicative in their impact on total, absolute survival and reproduction. Thus the unavoidable conclusion is that relatively small amounts of inbreeding can do tremendous damage to the reproductive potential and productivity of a fish stock.

Expected inbreeding depression is related to the current inbred state of a population. For example, a very small and isolated lake population of a certain fish species may be highly inbred because of its inherent demographic structure. The inbreeding of such a fish would not be expected to display an inbreeding depression as large as from a previously outcrossed group.

In some breeding systems inbreeding is the goal and its associated "depression" is an undesirable but expected result for which compensatory breeding programmes can be utilized. Using quantitative methods of Nace et al. and assuming an average recombination frequency of about 0.1, Nagy et al. estimate that one gynogenetic generation is equal to 10–12 generations of full-sib mating. In many other breeding and brood stock selection programmes, practical logistic and economic management considerations can and have resulted in the inadvertent selection of brood stock in a way which causes close inbreeding. Unfortunately, once it has occurred, inbreeding depression is not reversible except by hybridization.

Monitoring and Measuring Genetic Variation

Technologies exist for a direct assessment of the genetic properties of a population. In some species the genetic basis of variation in some visible characters (e.g., colour patterns) can be established by breeding experiments, and the characters, or phenotypes, can be used to directly assess gene frequencies in populations. Traits such as colour patterns are typically controlled by a small number of genes (one to three). As such, those genes may not be representative of either overall genetic variation or population structure, but probably reflect fine-scale ecological or social structuring such as family or age-class recognition. Although these characters can be conveniently assessed in a population, they will too often give misleading or unrepresentative information on genetic variation.

Electrophoresis of proteins has been widely applied for the direct study of genetic variation in fish populations. The importance of electrophoresis to the study of fish genetics resides in the ability to directly estimate genetic relationships from its results, and also because variation of electrophoretically detectable genes is often correlated with variation of other genes. To the extent that such a correlation is widespread among fish species, electrophoretic variation can be a general estimator of genetic variation.

The "state of the art" is to use electrophoretic techniques to analyze genetic variation in natural populations because, among other things, electrophoretic variation is "taxonomically congruent" with morphological variation in interpreting phylogenetic and evolutionary relationships . However the use of electrophoretic variation analysis requires some qualifications regarding aquaculture. At this time there is no direct evidence in the literature to indicate that either qualitative or quantitative allozymic variation is indicative of potential economic performance in such characters as food conversion rates, tolerance to temperature extremes, low dissolved oxygen tension, etc. Because of this, caution must be exercised in using the level of allozymic variation as the only criterion for choosing stocks for desirable physiological, nutritional, or other related production performances. There is some evidence of a correlation between electrophoretic variation, meristic variation, and developmental stability in nature . It is theoretically possible to develop the use of qualitative variation as predictors or indicators of quantitative performance in laboratory or hatchery populations but this involves complex breeding systems probably beyond the practical scope of most aquaculture breeders .

There are situations in which it may be desirable to electrophoretically estimate genomic variability and then to use these data as a base-line for comparing the genetic effects of a particular pattern of stock breeding or exploitation. For example, when exploitation of a species can be anticipated, base-line information on the genetic variation of pre-exploited stocks would be desirable. This would allow some direct assessment of the genetic consequences of exploitation by continuous monitoring of exploited stocks, or populations.

When re-introduction of a locally extinct population is contemplated, earlier base-line information might allow a closer matching of the introduced fish to the original population. Proper genetic matching would increase the likelihood of successful re-introduction. When base-line data is not available (the usual case), direct genetic assessment of the potential parental stocks for re-introduction allows an intelligent choice of stocks for introduction. Other things being equal, populations of maximum electrophoretic variation should be selected for introduction because this probably increases the likelihood of evolutionary adaptation to a novel environment.

A similar genetic monitoring of cultured species would be desirable to assess genetic changes that result from a particular culture scheme. It may be important to maximize outbreeding in a stock, in which case electrophoretic variation would be an important tool for monitoring the breeding programme.

There are certain dangers associated with the absence of genetic monitoring. In the south-western United States, a major breeding programme was undertaken some years ago in order to produce sterile males of the screw worm fly for introduction into the wild populations . During the breeding programme, a particular allele of the electrophoretically detectable gene dglycerophosphate dehydrogenase was accidently selected probably by natural selection in the culture population. This enzyme is important in flight metabolism. The properties of the selected allele militated against flight in the wild of the introduced, sterile males. Knowledge of the biochemical properties of the alleles and electrophoretic monitoring would have avoided this unfortunate situation.

We do not wish to imply or recommend that every exploited fish or every cultured stock be monitored in this way. Not only would this be expensive, but unless such studies are properly designed and controlled, the data are not likely to be very useful. We do, however, suggest that several carefully designed monitoring programmes be set up, and that these be coordinated to maximize their utility.

Population Genetic Structure in Relation to Exploitation and Extinction

Fishes probably surpass all vertebrate groups in their variety of social structures and kinds of life histories. It is not surprising therefore that some controversy has arisen regarding the significance of such variables as population structure, dispersal, and genetic drift, particularly with regard to genetic integrity of populations. At one extreme there are species like the American eel in which the adult population is spread out over thousands of kilometres, yet this species apparently verges on being a single random breeding (panmictic) population. At the other extreme there are hundreds of species which are territorial, which have demersal eggs with parental protection and which have very limited vagility. In species of this latter type with a localized and fragmented type of population structure, the neighbourhood size or the local population could be as low as 100, and there may be very limited gene flow between the local populations. As a consequence of this diversity, it is hazardous to generalize about demographic, georgaphic and genetic structural characteristics of fish. Each species must be examined as a unique case, even recognizing the possibility for intra-specific variation in population structure.

A good knowledge of the population structure in the management of fisheries cannot be exaggerated, whether the purpose of the management is exploitation or preservation or both (as should often be the case). Only when stocks are properly defined can the fishery be managed optimally. For example, one might conceivably decide to artificially enhance a pink salmon fishery by introducing fry from a hatchery. But unless one knew that odd and even year pink salmon were genetically distinct populations, the resulting hybridization could cause significant genetic change and a decrease in fitness in both odd and even year stocks.

Even very closely related sympatric species can have very different population structures, and it is dangerous to generalize from one to the next . In the rainbow trout in the Pacific northwest, the genetic data differentiate the populations into eastern and western groups, the major division coinciding with the crest of the Cascade Mountains. Many early workers had concluded that the principal basis for genetic separation of rainbow trout populations was anadromy and time of return to fresh water. Allendorf and Utter, however, emphasize that the taxonomic units based on electrophoretic data correspond to geographic groupings rather than to the above behavioural characteristics. They emphasized the importance of glacial events in dividing these populations historically.

In coho salmon, however, glacial events are not considered to have played a major role in the present subdivision of the populations in the Pacific northwest. The discontinuous distribution of (certain) transferring alleles among coho salmon populations cannot be directly explained on the basis of glacial events.

In the chinook salmon the coastal populations appear to be genetically distinct from the inland populations in both Oregon and Washington for the fall-run salmon, but the line of geographical demarcation is different from that of rainbow trout. This information has significant management potential because it could allow the determination of the major areas of origin of ocean caught fish.

Hence, among three closely related species in the same region, there is absolutely no correspondence in the geographic distribution of racially distinct populations, i.e., the geographic barriers that separate sub-populations in one salmonid species are not relevant in another. Management generalizations based on the distribution of populations in one species could prove disastrous if applied to other species.

Population structure is thus a useful guide for a priori priority ranking with regard to genetic resource preservation, at least with regard to the extinction potential of local populations. Geographic range alone is quite useful. The majority of species which reproduce in estuaries, river systems in the temperate zones, and in coastal pelagic zones are relatively widely distributed and fairly numerous. Species residing in tropical floodplain rivers, extreme environments such as shallow desert lakes or salt lakes may be far less numerous and typically have rather limited geographic ranges.

Many such "local" species are relatively vulnerable to habitat disruption. Both from the standpoint of biological conservation and from that of genetic resource preservation. Species which are widely distributed and relatively numerous warrant less attention than do the species with very limited distribution and small population sizes. We do not mean to imply that genetic depletion is only a threat to local endemics.

The potential for genetic depletion can also be quite high for a single reproductive unit of a broadly distributed species. The obvious distinction between the two cases is that the extinction of the endemic cannot be reversed, whereas the recolonization of a reproductive habitat upon loss of the more broadly distributed species is possible. For example, the Japanese sardine Sardinops melanostictafollowing a massive population collapse and range contraction, has recolonized the Japan Sea from refugia on Japan's east coast.

In the latter case, the genetic characteristics of the replacement population should be a relevant concern. For example, failure to consider the particular ecological adaptations of replacement or enhancement of stock has caused severe management problems in bob-white quail . In Scandinavia some introductions of Arctic charr and whitefish into lakes already populated by conspecifics has had deleterious consequences .

It is a moot point whether a naturally or artificially recolonized stock becomes as well adapted as the original stock. The point is to maximize the chances of a successful restocking or recolonization by taking into account all of the relevant genetic and ecological variables.

Criteria for Minimum Population Sizes

Criteria and Time Scale

The establishment of minimum viable population sizes is one of the principal goals of preservation genetics. In arriving at such minimum sizes it is necessary to consider all aspects of the biology of the species involved, not just the genetic. Other important criteria would include the demography and life history of the species, and certain ecological variables, for example, the probability and severity of catastrophe.

Very little in general can be said about the latter subject because the nature and consequences of catastrophe are highly dependent on the life history of the species and particularly on the kind of environment in which it lives. For instance, fishes living in shallow bodies of water in regions subject to extreme drought are likely to have a high probability of extinction. Also, they may go through severe bottlenecks with significant frequency. At the opposite extreme, deep sea species are unlikely to be exposed to physical events which bring about a collapse in population size.

Given sufficient demographic information, it is sometimes possible to produce estimates of minimum viable population sizes for a species. But even when the demographic information is at hand, the genetic approach is relevant and could be the dominating consideration, assuming that the minimum sizes based on a genetic criterion are smaller than those based on demographic or ecological criteria.

The "time scale of survival" is a useful device for structuring a discussion of preserving genetic variation. Somewhat arbitrarily, there are three problems or issues:

1. a short-term issue is immediate fitness - the maintenance of vigour and fecundity during an interim holding operation, usually in an artifical environment, such as when breeding domesticated or semi-domesticated fish stocks.
2. the long-term issue is adaptation - the persistance of vigour and evolutionary adapatability of a population in the face of a changing environment;
3. the third issue is evolution in the broadest sense, i.e., speciation, or the creation of evolutionary novelty . For our purposes, the first and second issues are the most relevant and the third is the least relevant.

From the above discussions on heterozygosity and inbreeding in both natural and captive populations, we can conclude that excessive loss of genetic variability, particularly inbreeding, can and must be avoided. The issue, however, is a quantitative one and the above qualitative discussions are not very helpful in providing specific guidelines to minimum viable population size. We must examine this point in greater detail.

Survival of Captive Populations

Captive populations will tend to be small and potentially subject to the deleterious effects of inbreeding. By trial and error, animal breeders have discovered the magnitude of inbreeding that can be tolerated by domestic animals before the lines begin to decline in fitness. (In discussing inbreeding, it is convenient to use the inbreeding coefficient, F, which is a quantitative measure of the magnitude of inbreeding. In a population that is totally outbreeding, F = 0. For a population that is totally inbreeding, F = 1.0.)

A general rule is that the per generation rate of inbreeding should not be higher than one to three percent . Higher rates of inbreeding fix deleterious recessive genes too rapidly for selection to eliminate them, and the vigour and fertility of the line decreases.

The lower inbreeding rate of one percent (F = 0.01) is preferred because:

1. poultry and mammal stocks have been partially purged of deleterious genes over the milennia which allows them to tolerate higher rates of inbreeding than wild outbreeding species;
2. animal breeders can safely ignore some inbreeding and random loss of genes. In constrast, conservationists wish to preserve the "wildtype".

How does this basic "one percent rule" translate into population size? The rate of loss per generation of heterozygosity due to inbreeding as measured by F is equal to $1/(2\,N_e)$, where N_e is the effective population size. ("Effective population size" is the size of an idealized population. The definition of N_e is cumbersome, but the population must have an equal sex-ratio and individuals must mate at random. A number of additional "ideal" characteristics could be stated. For our purposes, it is important to note that in practice N_e is nearly always smaller than the actual number of breeding individuals.) Thus, N_e must equal at least 50 if the inbreeding rate is to be kept below the one percent level.

However, even when F = 0.01, the loss of genetic variation is appreciable after a few generations, and a gradual attrition of genetic variation cannot be prevented. Eventually, the population will become virtually homozygous, the time depending on N_e. This is why the one percent rule must be viewed as short-term criterion. A population held in check at N_e = 50, will lose about one-fourth of its genetic variation after 20 to 30 generations, and along with it, much of its capacity to adapt to changing conditions. Thus, if it is desired to maintain a particular stock for longer than this, it will be necessary to increase its N_e. A rough rule of thumb is that G is approximately equal to N_e, G being the number of generations the stock is likely to retain its fitness at a relatively high level.

The above information is necessary but not a sufficient basis for the conservationist or the fisheries' biologist to conserve short-term fitness, or to maintain short-term fitness in captive populations of fish. The reason is that the effective population size is not a simple phenomenon and is affected greatly by variation in sex ratio, population size through time, by a non-random distribution of progeny among families, and other aspects of the breeding system. To the extent that any of these effects occur, a larger absolute population size must be maintained to achieve a desired N_e.

Population Influences on Effective Population Size

When populations decline or "crash", the survivors constitute a genetic "bottleneck" in the history and evolution of the population. Any deviation in the genetic makeup of these survivors from the gene pool of the original population will be reflected in future generations. More particularly, if the progenitor's gene pool is less diverse than that which existed in the original population, future generations will have a corresponding deficit in genetic diversity.

If the minimum population size is very small, due either to normal fluctuations or to an environmental changes or catastrophes, it is tantamount to squeezing the genetic variability of the source population through a very narrow channel and eliminating a significant amount of this variability. Bottlenecks inevitably accompany the establishment of a captive stock for breeding purposes.

Prevention or further genetic erosion or a recovery to the original level of genetic variation depends greatly on how fast the population grows to a moderate size or several hundred or more. If a preserved population is subject to fluctuations in numbers (as it most probably will be), the influence of the minimum absolute size on effective population size is more relevant to preservation of genetic diversity than the average absolute size.

The loss of genetic variability concomittant with the bottleneck event has both qualitative and quantitative aspects. Qualitatively, specific alleles may be lost and if they are lost it is very unlikely that they will be replaced by mutation as long as the population remains small. Quantitatively, the variability for specific traits will be reduced and the mathematics in the loss of the variance of quantitative traits have been described by Falconer and others. The qualitative effect is usually greater than the quantitative one; that is, the loss of alleles, especially low-frequency alleles, is much greater than is the loss of genetic variance per se. Incidentally, several workers have pointed out that the number of founders in a colony, so long as it is greater than about five individuals, is not nearly as important as the long-term maintenance size of the colony. That is, a single bottleneck event followed by rapid growth to a large size, say $2N_e$ greater than 500, does relatively little damage, compared, that is, to a chronically small N_e.

We know from experience with resistance to pesticides in anthropods that some alleles that occur at very low frequency in natural populations (and which are likely to be lost during a bottleneck) can be very important. Such alleles can mean the difference between survival and extinction. The same thing probably applies to resistance genes in general. Therefore, it would be expected that populations of fish passing through bottlenecks might not be noticeably affected until a disease epidemic swept through the population. Only then would the loss of these resistance genes be detectable.

Several other factors determine N_e. Among these are the sex ratio; N_e is lowered by deviations from an equal sex ratio. Another of the characteristics of a genetically ideal population is that the number of progeny are randomly distributed among families. When this condition does not hold, for example when the reproductive output of a few families is especially great, N_e will be lowered. It is incumbent on persons dealing with captive populations, for purposes of either preservation or culturing, to be aware of these effects and to maintain N_e at a level that will maintain the fitness of their stocks. In some cases, it will be desirable to consult with a population geneticist, especially if the breeder is in doubt about the estimation of N_e.

The Maintenance of Long-term Fitness

Long-term preservation requires rather large population sizes, large enough so that an equilibrium will be maintained between the loss of genetic variability due to drift and selection and its generation from mutation. When $2N_e$ is a large number, say greater than 500 or 1 000, the effect of drift will be negligible compared to that of weak selection. When $2N_e$ is small, say less than 100, the randomization of gene frequencies between generations will not only fix many loci, it will also counteract all but the strongest deterministic forces, particularly directional selection, thus, to a large degree precluding adaption by natural selection.

The consequences of small population sizes might be thought to be irrelevant when considering the genetics of fishes in an aquatic reserve, or similar ecosystem. However, there will always be some populations in a natural ecosystem, particularly large predators, which have quite small numbers (such as groupers on an atoll, pike or muskies in a lake). The loss of such keystone predators from a natural community can have serious effects on the diversity of prey species, as has been documented by many workers, particularly with marine invertebrate systems .

Franklin argues that a minimum effective size of 500 is needed to preserve useful genetic variation, because:

1. The relevant phenotypic traits in conservation are quantitative (polygenic). For such traits the average effect of a gene is small, and most of the genetic variation is additive.
2. Weak directional or stabilizing selection does not erode additive genetic variation at a significant rate;
3. The significant evolutionary forces, therefore, are mutation, and genetic drift. That is, if a population is below some threshold size, it loses variation by drift at a faster rate than it gains variation by mutation.

Franklin derives his number from the work of Lande on bristle number variation in Drosophila. The evidence is meagre but Franklin believes his number (500) is about the right order of magnitude. Simple theory also yields this number as mentioned above.

It is necessary to caution again that the employment of any number is subject to all the same qualifications given in the preceding section for short-term preservation, namely that any effective size translates into a much larger number of breeding adults, when dealing with real, not ideal, populations. In addition, this recommendation ignores genetic differences between species. Conservation questions of this kind must also be considered on a case by case basis.

Components of Genetic Impoverishment in Fish Populations

Human activities can cause the erosion of genetic variation and the extinction of fish via several mechanisms. These include:

1. Pollution and other environmental changes that stress a population and cause differential mortality, extinction, or both;

2. Fishing pressure which can favour some genotypes over others;
3. Artificial selection and domestication which can result in conscious or unconscious inbreeding and genetic impoverishment;
4. The introduction of exotic species and diseases.

Effects of Pollution and Other Stress Factors

Pollution of different kinds (e.g., eutrophication, toxins such as mercury, DDT, PCB) have had disastrous effects on many fish communities. In most western countries toxins in single fish species have made them unfit for human consumption. In some cases legislation has been successful in reversing this situation. For example, a large part of Lake Vänern, Sweden (the third largest lake in Europe) has been blacklisted, but the situation has improved by changing the water purification procedures of the pulp mills.

The paramount problem in Scandinavia and elsewhere has been airborne acid rains (SO_2) from industries using fossil fuels as a source of energy. For example, large parts of southern Norway and southwestern Sweden have become almost devoid of fish due to acid rains originating from Great Britain and the Federal Republic of Germany. Species of fish differ in their response to acidification. Roach (Leuciscus rutilus) and most salmonids are extremely sensitive to decreasing pH. The addition of lime has been tried as an "artificial breathing" agent (1 million Swedish Kroner a year in Sweden), but it is obvious that this problem cannot be solved without some kind of international negotiation. Several charr populations have recently become extinct due to their sensitivity. The primary effect of acidification is apparently on the early reproductive stages.

Pollution by nutrients (especially PO_4) from human communities has been most disastrous to densely populated areas in temperate, sub-tropic and tropic areas. Despite this, in arctic or alpine areas with extremely oligotrophic environments, eutrophication by adding nutrients has been promoted as a method of increasing productivity. This is now being studied experimentally in Scandinavia as well as in Canada.

In some cases, "nutritional pollution" has been successfully treated. In all these cases, the construction of purification plants has resulted in significant improvement in water quality. In Lake Washington (U.S.A.), the spawning run of Pacific salmon species is now restored and natural spawning is occurring. Likewise, there is now a natural recolonization of native species in the lower Thames in England. The construction of purification plants in all communities around Lake Malaren in Sweden has made it possible to reintroduce Baltic salmon and sea-run brown trout into a stream that discharges through Stockholm.

Hydroelectric power development has also affected the fish populations in impounded lakes and the river stretches between the power plants. An important observation in Scandinavian waters is that there are sub-populations of salmonids that are affected in different ways. For instance, brown trout is often represented by different populations in different sites within a stream, apparently with little interbreeding. During the construction of dams the populations spawning in the outlet area become extinct. Arctic charr and whitefish are often represented by two or more discrete populations which are affected in different ways. On the whole, pelagic populations are less affected than benthic ones.

There are only a few rivers remaining in Sweden which support natural spawning of salmon. In the Baltic the majority of salmon originate from Swedish hatcheries. The smolts released have been carefully selected, so as to represent the original native stocks. There is evidence for genetic change, however, mainly toward smaller size and earlier homing (earlier maturity), this is probably an effect of overfishing.

The landlocked salmon of Lake Vänern are represented by two sub-populations which do not seem to interbreed. These are: the "Klarälven" salmon, which spend three years in the River Klaralven and return for spawning after three to four years, and the "Gullspångsälven" salmon which spend two years in the River Gullspångsälven and return from the lake after four to five years. The growth rates of the two populations are strikingly different.

Tagging experiments with reared smolt have shown that after 40 months in the lake the salmon from Gullspångsälven attain a weight of about 5 kg, whereas the salmon from Klaralven reach only 2.5 kg. They are also segregated spatially during their life in the lake, suggesting an inherent capacity for both growth and habitat selection.

Both populations are endangered, mainly by hydroelectric development. The fastgrowing population from the River Gullspångsälven is very close to extinction, but efforts are being made to save it, both by hatchery operations and introductions into other lakes.

Effects of Fishing

Overfishing for Baltic salmon has resulted in slower growth, smaller size at maturity and early homing. Very similar results were obtained by Gwahaba in the case of Tilapia nilotica in Lake George, Uganda. He claimed that the restricted size of nursery zones, fish killing storms and predation have been factors in the decreased propagation, but an increased commercial fishery appeared to be the main reason for the changes.

Turner provides another example of ecosystem change by intensive fishing on Lake Malawi cichlids. Whereas at the beginning the catch was dominated by large species, small species now predominate. This example dramatically shows the effects of "overfishing" on an entire fish community. It is predictable that in the case of continuous fishing with nets of unsuitable meshes, the larger species may be threatened with extinction, in part, because the more abundant smaller species can outcompete the small stages of the former large ones.

Similar observations were made comparing discrete populations of Arctic charr which are subjected to different exploitation rates..

Effects of Exotic Fish Species

Active introduction of food fish

Since the early 1970's much information has accumulated on the serious consequences of introductions of exotic (foreign) fish on native ichthyofauna. For example, many formerly fishless, deglaciated lakes were stocked by the Lappish people and European settlers in Scandinavia and in North America by the Indians and European settlers. As hatchery technology has improved, the practice has spread.

The rate of such introductions has greatly increased in recent years. According to Everett rainbow trout (Salmo gairdneri) was introduced to the Lake Titicaca Basin in Peru and Bolivia in 1942. In the 1950's the following trout species were also introduced: S. trutta, Salvelinus fontinalis, S. namaycush and the atherinid Basilichthys bonariensis . By 1972 two of the most valuable (largest) species of the endemic genus Orestias in Lake Titicaca became practically extinct by predation or food competition from the trout species. Sporozoon parasites which were introduced passively together with the exotic species account for the dramatic population declines in the majority of endemic species of Orestias .

Another well known example is that of Lake Lanao on the island of Mindanao in the Philippines. According to Villaluz (1966: cited by Frey, 1969) personnel associated with the State University of Marawi City on Mindanao introduced Clarias batrachus (Siluridae),Ophiocephalus striatus (Ophiocephalidae) and Tilapia mossambica (Cichlidae). Even more serious was the accidental introduction ofGlossogobius giurus (Gobiidae) at the same time. According to Frey at least some benthic and pelagic species of the endemic genusBarbodes (Puntius) have become rare as evidenced by their virtual disappearance from the local fish markets.

Both these species groups, the endemic Orestias in Lake Titicaca, as well as the Barbodes in Lake Lanao, were the principal economic and nutrition resources of the local residents. These cannot be replaced by the introduced species because the introduced species are not readily accepted by the local population. The effects on the genetic resources of the endemic species ofOrestias or Barbodes are irreversible. Some species are evidently extinct and even those which may have survived have probably been diminished to one or a few small, disjunct populations.

Introductions of sport fishing and ornamental fish

Game fishermen and aquarists have been responsible for numerous cases of introductions which have led to detrimental changes in the native fauna. Studies in Scotland and Scandinavia have shown that native species may become extinct by ecological competition with exotic ones. For example, Svardson described the effect of introductions of species of whitefish (Coregonus) and consequent extinction of the Salvelinus alpinus complex in Scandinavia.

Hybridization between native and related, introduced species also has serious consequences on genetic resources. In the second half of the 19th century different trout species were introduced into California, confusing the already complex situation in the native trout fauna. Similar events have been reported by Moyle for species of chub (Gila) and sticklebacks (Gasterosteus). The most frequent result is swamping of the genome of the native species. The large number of hybridization events reported from Florida were caused by aquarists or the aquarium fish industry . Twenty exotic species and five hybrid populations have established themselves as breeding populations.

Introductions for weed and insect control

Moyle reported on fish introductions in California for weed and insect control. In addition to several killifish and two species of Tilapia, the mississippi silverside (Menidia audous, Atherinidae) was introduced illegally in 1967. Since then it has become the most abundant and widespread species. The Tilapia as well as Menidia have affected the relative abundance of the different fish and the total number of zooplankton organisms to which native fish were highly adapted, although it is difficult to predict the ultimate consequences for the native species.

Both food competition and predation on the larvae by exotics of native species have often led to extinction. Within less than a decade, mosquito fishes (Gambusia, Mollienisia) introduced into southern European countries and North Africa were responsible for the decline of the native Aphanius sp. population (Cyprinodontidae). Similarly, Cyprinodon in the Colorado River system is endangered by direct competition of introduced live-bearing forms.

Accidental introductions

Accidental introductions have occurred in several ways. Among these are uncleaned gillnets ("egg pollution") and the escape of bait . Another danger is the escape of stocked and exotic species from aquaculture ponds, for instance by flooding during rainy seasons. Introductions have also occurred via canals or tunnels used for shipping and power plants. The most famous case is that of the landlocked sea lamprey (Petromyzon marinus) which entered the Upper Great Lakes through the Welland Canal. Similar movements have occurred into the Mediterranean Sea since the opening of the Suez Canal.

The biological consequences of introductions

Many introductions have been economically or aesthetically successful. For example, most species of North American salmon, as well as the black bass, have been introduced into European fresh waters, and several Oncorhynchus species have been introduced into the Northern Russian marine waters and to the Baltic. Kokanee (Oncorhynchus nerka) has been introduced into Scandinavian freshwaters. Although most of these introductions have failed, black bass in Germany and rainbow trout in Great Britain have been successfully established. Brown trout and carp from Europe have been introduced into North America, New Zealand and most alpine areas where

Europeans have settled. The introductions of lake trout (Salvelinus namaycush), which began in Fennoscandia in the 1960's, seem to have been successful in some Swedish and Finnish waters.

Little is known about the impact of these introductions on the genomes of the species, but where intense culturing and domestication is practised, for instance in rainbow trout, genetic change is inevitable. The same should be true for brown trout in Europe and elsewhere.

A very intense study on the effects of transplanting Coregonus and Salvelinus species has been carried out in Sweden and Italy. Briefly it has been shown that introductions of "exotics" have four possible results:

1. extinction of "ecological homologues". The most drastic example has been the extinction of Arctic charr (Salvelinus alpinuscomplex) by introductions of certain whitefish stocks;
2. hybridization with concomitantly profound effects of the genetics of the original fish population;
3. failure of the introduction, in part, because of competition from established resident species;
4. coexistance, which means that the introduced species has found a "vacant niche" in the community, with an interactive niche segregation as an obvious result.

Introductions of exotic prey species is now a common practice, especially in North America and Scandinavia. In Canada and Sweden, for instance, introductions of glacial relicts (especially the crustacean Mysis relicta) have been practised for several decades . Such introductions have on the whole seemed successful. In many cases, however, a strong impact on the native zooplankton community has been observed.

Effects of Artificial Selection

Artificial selection is usually associated with desirable genetic changes. However, in many cases the outcomes of artificial selection may be neither desirable nor predicted. This is especially true when brood stock selection and management lead to loss of genetic variation. By their very nature, specific examples of genetic impoverishment due to such practices are not generally available. Nevertheless, there is sufficient background knowledge to warrant caution. For example, genetically based performance under one set of conditions (i.e., hatchery) may not be correlated with performance under a different set of conditions (stream, lake, natural area). If the goal is to release stock in a different environment from that where they are bred, then the brood stock selection practices must be designed to avoid unconscious selection and inbreeding.

In this regard, the accessibility of brood stock to human managers does not ensure its genetic superiority. For example, it has been shown that large carp that are chosen from a population of identically aged individuals do not necessarily represent superior genotypes . Such individuals may represent an exposure to a favourable set of environmental conditions which magnify a small, initial, non-genetic, size differential into a large one. The use of a brood stock composed of such individuals would lead to little or no genetic progress, and could, if the number of selected individuals is small, lead to inbreeding and genetic drift. Moreover, large individuals could represent behaviourally aggressive phenotypes under genetic control. The choice of such individuals could lead to an undesirable general level of agressiveness in the population.

In some cases brood stock selected from culture units containing mixed year classes (as is probable in undrained tropical ponds) probably represent a mixture of both superior and inferior genotypes. Consequently little genetic progress, and even erosion, can be expected by the haphazard selection of brood stock. For example, in the case where only gravid females from undrained, net harvested ponds are used for brood stock, they may represent both large young animals (superior genotype) and large older ones (inferior genotype). In addition, it may be that genetically inferior (i.e., slow growing but "net avoiding" males) are reproducing more often with these females.

It is in part for reasons such as these that Tilapia pond culture has failed in some tropical areas. In harvesting Tilapia ponds it is not uncommon to select the large fish and to return the small ones. This procedure favours the reproduction of small and slow-growing fish, and it has been shown experimentally that such selection can produce a genetic shift in the population toward genetically smaller (and less desirable) fish. The problem in Tilapia is compounded by mouth breeding and defence of young until they reach a size safe from cannibalism.

The result is overpopulated ponds with stunted individuals. Cannibalism can be genetically and economically advantageous in some aquaculture systems.

Domestication

Domestication is defined as a genetic selection process mediated by geographic and reproductive isolation, inbreeding, and small population size to produce profound and desirable evolutionary changes in a genetic stock. Domestication presumably involves at least the ability to live under artificial conditions through most parts of the life cycle. It further would usually be expected to include genetic adaptation such as to crowding, handling, and artificial diets. Genetic improvement, in the sense of enhanced growth, modification of body form and some loss of fear-flight behaviour are also usually assumed. The anthropological aspects of this are well documented , but not much scientific scrutiny has been directed at the domestication process itself because of the limited opportunity to study, document, and control the phenotypic and genotypic changes that occur in the domestication of livestock, poultry, and companion animals. The opposite is true in aquaculture.

There are probably a number of reasons why man has not domesticated as many fishes as he has mammals, birds and plants. Because of the difficulties of transporting live fish from place to place over land, domestication could only have proceeded in sedentary cultures. There is indeed evidence of fish culture in Sumeria contemporaneous with very early stages of agricultural development, pre-dating the reported beginnings of aquaculture in China in 600 B.C.

It is also likely that attempts to apply the criteria for selection learned in animal and plant breeding to fish or to cold-blooded vertebrates generally were counter-productive. Whatever the reason, and with the notable exception of the goldfish, ornamental carp, and perhaps the common carp, few fish could be considered domesticated even though some strains of trout, for example, are clearly much more adapted to hatchery conditions than their wild counterparts .

Culture conditions themselves, along with the conscious selection imposed by man, will foster domestication. By and large the outcomes of this process are not only desired but predictable. Workers might be cautioned, however, that, history rarely reports failures in such attempts. In this regard researchers, culturists and breeders should develop the perspective that these efforts represent, in many cases, the initial steps of incipient domestication of a new animal group. Their scientific scrutiny should be designed to monitor and document this process.

It is recognized that in view of the inevitable operation of natural selection in culture environments that propagation without domestication is, perhaps, impossible. However, in some circumstances breeding without genetic change is the goal, as in the case of stock enhancement programmes of natural fisheries, in which it is desirable to preserve the inherent, undomesticated genotypes.

Thus care must be taken to avoid the development of traits that have historically charracterized domestication (e.g., loss of awareness of predators, agressivity, irritability). Such losses may prove useful in brood stock management because these traits result in increased accessibility and tractability to human management. However, development of such traits will be counterproductive to the goals of stock enhancement of natural fisheries.

In some cases of stock enhancement, behavioural genetic change under domestication is useful and desirable. For example, impounding of areas in coastal reclamation projects or inland reservoirs can create vast areas of semi-natural aquatic habitat useful for aquaculture. In such situations, the loss of migratory and other inappropriate social behaviours may be desirable.

It might be expected that the scientific knowledge now available will result in relatively rapid domestication of aquatic species as suggested by Moav et al. . Techniques such as gynogenesis are likely to facilitate the domestication process dramatically. The normally high fecundity of fish species has been cited as contributing to rapid domestication. Whether or not this happens, it is surely essential, as was pointed out in the recommendations of the 1976 FAO Technical Conference on Aquaculture held in Kyoto, that fish breeders provide very careful documentation of the breeding history of strains undergoing domestication.

Preservation of Genetic Resources in Closely Managed Fish Populations

Artificial Selection of Broodstock

Artificial selection of broodstock is broadly defined as the conscious selection of an individual or genetic group for the purpose of providing progeny for culture or stock enhancement. This selection can be based upon the performance of an individual (individual selection), of its progeny (progeny selection), of its ancestors (pedigree selection), of its contemporary relatives or on the potential of a useful combination with a parent of another genotype (hybridization selection). In all cases selection is practised in order to improve the performance of the progeny group above that of a progeny group from unselected parents.

There are many selection methods available by both the professional and non-professional aquabreeder . The proper application of the methods in a breeding programme can lead to useful and desired economic, aesthetic and ecological outcomes while still maintaining the genetic variability in the population. However, improper application of artificial selection or its application without sufficient background knowledge can lead to a deterioration of the genetic base.

Useful outcomes, both experimental and economic have been documented for artificial selection in aquaculture. In these particular cases, undesirable outcomes have either been economically inconsequential or predicted (as in the case of experimental systems). In many cases, however, there have been undesirable results. This happens when simple brood stock selection and management per se leads to artificial selection.

The use of controlled mass rearing hatchery techniques dramatically increases the survival of larvae, post larvae and fry above that realized in nature. In theory, higher survival rates should increase the genetic variability available for natural and artificial selection, because they can "open up" the genotype of a genetic group in so far as new genetic combinations are allowed to survive. In the long term, however, culture conditions impose their own "natural" selection on the genotype. This may or may not lead to a desirable outcome and must be monitored carefully.

Propagation of fish stocks in hatcheries for the establishment of a selected strain takes anywhere from ten to 50 generations. Fish breeders, therefore, might be tempted to maintain a minimum number of individuals. It is thus essential that they carefully weigh the costs of decreasing fitness and genetic variability against the economic advantages of maintaining such small numbers of individuals.

Fitness is expected to decrease so long as the effective population size of a brood stock is less than 50, and genetic variation in quantitative traits is expected to leak away unless the size is on the order of 500. If breeders choose to ignore these guidelines, they should do so in the knowledge that the genetic health to utility of their lines could be endangered.

Artificial Methods of Reproduction

Methods of artificial propagation, including the use of hormones, in vitro fertilization and development, are in wide use in aquaculture. For some species artificial propagation is necessary for life cycle control. In others, such propagation techniques are more efficient than natural breeding methods. The ways in which artificial propagation can affect the maintenance and study of genetic variation include:

1. Increase of genetic variability and genetic understanding. Numerous broods from different breeding pairs and from single maternal and multiple paternal (and visa versa) crosses allow not only the avoidance of inbreeding and bottleneck effects but the integration of selection operations with proper experimental designs. As pointed out in the report of the FAO Ad hoc Working Party on Genetic Resources of Fish this integration allows both an improvement of stocks and a contribution to the knowledge of the genetic control of the production characters.
2. Aids in the conservation of stocks, strains, species and other genetic groups. Artificial propagation and life cycle control allows the conservation aqua-breeder to maintain valuable or potentially valuable stocks for current and future use.
3. Allows the development and use of specialized breeding methods. Gynogenetic procedures have been developed with the use of artificial propagation techniques. These procedures are used to create and maintain highly inbred lines which can be used to create and preserve useful homozygous genotypes. By maintaining numerous such homozygous lines in a population, overall genetic variability can be preserved despite the fact that each line represents severely reduced variability. Crosses of the lines can create heterozygous variability which may be economically useful because of hybrid vigour and/or because of the favourable non-heterotic combination of genotypes. Heterosis of 30–40 percent above that of the homozygous, gynogenetic lines can be expected in crosses of these lines.
4. Allows the efficient maintenance of effective population size. Maintaining a minimum effective population size is essential for the conservation of genetic variability. Under natural conditions effective population size can vary widely. Through the use of artificial

propagation methods man can control the effective population size most simply through the control of the numbers of breeding males and females.

5. Aids in interspecific hybridization. Normally, behavioural and anatomical differences preclude interspecific or intrageneric crosses between species which otherwise are reproductively compatible at least for the creation of F_1 progeny. The use of artificial propagation techniques, especially those involving in vitro manipulation of gametes and in vitro culture of developing embryos in species which have maternal brooding, would greatly aid in developing interspecific hybrids.

Hybridization and Heterosis

Hybridization (or cross breeding) can be divided into two categories: (1) intraspecific: crosses between strains, stocks, land races, geographic populations within a species; (2) interspecific: crosses between species. Hybridization is practised to achieve either of two favourable outcomes. These are: (1) heterosis or hybrid vigour, which is defined in a broad sense as an increased performance or value (to the aquabreeder) of progeny above the average of the parental performances or value, and (2) non-heterotic effects which is the performance of the progeny as the result of simple combination of parental genotypes.

Specific hybridization is useful in producing a wide variety of new genetic combinations. Improvements in production above that attained for "land race" strains have been realized in extensive and intensive fish farming. The expected heterotic effect can be increased using highly inbred parental lines. The desired outcome is the homozygosity of useful genes in a highly standardized and uniform product.

Interspecific hybridization is used not only to search for heterotic effects but also to search for favourable combinations of genotypes that control traits and performances that do not vary within a species to a large degree. For example, new kinds of social and feeding behaviour, better adaptation to environmental extremes in natural and controlled systems, and better adaptation to new husbandry systems can be realized by interspecific crosses. In addition, interspecific hybridization can be used to create desirable sterile or monosex progeny groups which also display favourable production performance. For example, a sterile triploid hybrid between the big head and grass carp has been developed . This sterile hybrid can be introduced to ecosystems for weed control with no danger of overpopulation or crossing with wild populations. The creation of monosex broods of Tilapia sp. brought about by interspecific hybridization is being intensely researched . As yet no one combination has consistently resulted in a monosex brood. However, crosses leading to highly skewed sex ratios are possible and are being utilized.

Interspecific hybridization is not rare in nature. However, to date most human attempts at interspecific hybridization in fishes have been directed at exploited Salmonid , Cyprinid , and Centrarchid fishes. However, in all fishes, unexploited species represent a vast repertoire of behavioural, physiological, nutritional and other adaptations which may prove useful in making larger impacts in genetic improvement than would be obtained through the use of intraspecific genetic manipulation. The development and use of artificial propagation and in vitro techniques will greatly aid in this endeavour.

Enhancement of stocks

There are numerous examples of activities based on the concept of augmentation of natural population production by introduction of hatchery raised, "cultured" juvenile fishes into natural systems. Varying levels of success have been achieved. In some cases, for example, the early attempts to augment cod (Gadus morrhua) production in the North Atlantic Ocean have not been obviously positive. In closed systems such as lakes, inland seas, and reservoirs, however, there have been obvious improvements in the productivity of various species.

It is apparent that natural fish stocks undergo genetic alteration through addition and deletion of genetic material. Although quite a lot of attention has been given to effects of competition between "wild" and hatchery raised fish, for example in salmonids, very little has been given to changes which may result to the genomes of the wild stock by mixture with artificially raised stocks. In some cases parental material is taken directly from the population into which the hatchery fish are subsequently released. In this case, the principle hazard would be the possible loss of genetic diversity through long-term inbreeding of the parental stock. Where there is relatively low rate of natural reproduction or where very few spawners are used to provide most of the recruitment, random loss of genetic variation (genetic drift) would increase. When, as is often the case, one hatchery is used to provide seed for stocking many lakes, the effects are much more likely to be important. Flick and Webster have reported higher survival of both wild and hybrid domestic strains of brook trout after release in small natural lakes compared to pure hatchery strains. In salmonids and carps, quite a number of distinct hatchery strains have been developed.

Genetic differences have been demonstrated between fish populations in different streams of the same river basin, between nearby lakes, and between sub-populations within the same body of water. There is, however, little direct evidence concerning the degree to which such variation among stocks affects "fitness" in the different habitats. Recent biochemical and physiological studies have shown, however, the existence of correlations between gene frequencies and environmental variables, such as temperature. Thus, local ecotypic adaptations of fish populations are to be expected and the genetic and ecological consequences of intraspecific transfers may be undesirable.

Cryopreservation of Genetic Resources

In order to maintain high levels of genetic variability in cultured fish, it would be necessary to maintain large numbers (i.e., hundreds) of breeders. This is expensive and often not feasible. The introduction of the cryopreservation of gametes which can be used when necessary, thereby reducing the genetic erosion that invariably results from inbreeding, allows one to freeze the sperm of numerous male donors from all possible strains.

Essentially the method involves stripping sperm, from males, diluting the sperm with appropriate extenders and life protectors, then freezing in liquid nitrogen or on dry ice followed by liquid nitrogen. The exact methodology is somewhat different for each species. Once the method has been refined for a given species, sperm from thousands of donor males can be kept for years thereby preserving the genetic resources for

(1) cultured fish stocks,

(2) unique natural populations, and/or

(3) species that are threatened.

At present methods for freezing fish eggs or embryos are not technically feasible but research in that area is being actively pursued.

Recommendations

Considering the broad terms of reference given to the Consultation, and the diversity of recommendations which arose from the discussion, the participants agreed to group their recommendations according to the differing audiences to which each was primarily addressed. The groups agreed upon are as follows:

1. International Organizations
2. Governments
3. Aquaculturists and Managers of Fisheries
4. Conservationists
5. Research Scientists

Much of the immediate impetus for holding the Consultation came from the continuing work of the United Nations Environment Programme (UNEP) to develop a global strategy for the conservation of genetic resources. For this reason the Consultation gave particular attention to activities that are needed to develop and implement a global strategy of conservation of fish genetic resources. Many of the following recommendations, in all categories, were formulated to contribute to the development of a global strategy, as proposed by UNEP, for long-term conservation. The first set of recommendations is particularly aimed at developing institutional support for programmes of genetic resources conservation.

The group also considered more immediate actions for effectively managing the present use of aquatic resources. Recommendations concerning such actions are mostly grouped under those directed toward governments and to individual fishery managers and aquaculturists.

Recommendations Primarily for International Organizations

1. FAO, through cooperative programmes with other international agencies, should promote "grass roots" awareness of genetic preservation concepts in the fishery/aquaculture communities at regional and national levels. These objectives could be met on the short term by development and distribution of training manuals and information documents which need to be followed up immediately by workshops.
2. Biological criteria for the design and management of aquatic reserves need to be defined from genetic, ecological and demographic principles.

A consultation of experts, with varied applied, experimental and theoretical backgrounds representing the critical areas of expertise, should be convened to define biological criteria for the design and management of aquatic reserves for preserving natural genetic stocks of fish and other aquatic organisms. The consultation should include genetic, ecological and demographic principles in their considerations.

3. It is recommended that international centres be established, preferably on a regional basis, where corps of experienced researchers would (1) assemble basic biological and genetic data on exploited and potentially exploitable fish stocks, and (2) disseminate such information through catalogues, newsletters, etc., in order to provide more direct mechanisms at the international level for the resolution of biologically important problems and conflicts, particularly regarding the exploitation of fish species. Special attention would be given to sub-species components of resources shared by two or more nations, and to species which may be considered for introduction into different drainage basins or across national boundaries.
4. Unesco and perhaps other international organizations such as FAO and IUCN, should consider establishment of a programme of education and training, at both national and regional levels, on genetic resource conservation/preservation in fish and other aquatic organisms. As a basis for its educational work, the programme should assemble baseline information (a) on the diversity and vulnerability of aquatic genetic resources, (b) on procedures for identifying vulnerable species and population, and (c) on appropriate methods assuring that information regarding vulnerability and direct threats comes to the attention of agencies competent to act.
5. Acid rain is at present a particularly serious threat to genetic resources of aquatic organisms and a problem requiring international action. FAO and other private and public international organizations should encourage governments to negotiate appropriate protocols for the control of the amount of sulphur that is discharged into the atmosphere.
6. The many international organizations for the regulation of fish stock exploitation (e.g., EIFAC, ICES, Great Lakes Fishery Commission, etc.) are encouraged in their efforts to prevent the extinction and genetic deterioration of valuable stocks.

Recommendations to Governments

1. Introductions of new species into aquatic systems has often had serious consequences on existing resources. Governments which do not now have mechanisms to ensure that an objective analysis of risks precedes the introduction of an aquatic organism into national waters should take immediate steps to establish such mechanisms. Genetic, behavioural and ecological data, as well as potential for introduction of disease, should be included in the risk analysis. In this connexion, governments should be aware that the probability of escape of cultivated aquatic species (even those kept only for research purposes) is so high that intent to confine imported aquatic animals does not obviate the need for such risk assessment.
2. Governments should consider urgently the establishment of fresh-water and marine reserves following principles which have been established for land reserves.
3. Governments should insist that the potential impacts of planned hydroelectric and irrigation, and other development projects upon fisheries and fish genetic resources be evaluated at the earliest stages of consideration of such projects to ensure that there is opportunity to examine appropriate alternatives.
4. Environmental alteration (by pollution, siltation and erosion, etc.) is generally a more important threat to the preservation of fish genetic resources than their direct exploitation. Governments wanting to protect these resources should make every effort to ensure that environmental damage to natural waters is minimized.

Recommendations to Aquaculturists and Fishery Managers

1. Fish breeders should be concerned for the continuing fitness of their stocks, and should maintain the effective population size, N_e, of the stocks at 50 or more for short-term breeding and culture programmes, and much more (ca. 500) for the protection of genetic variability within lines. Inbreeding techniques should be used only for specific genetic goals and only when strict genetic control is possible and can be used in conjunction with other selection programmes.
2. Research aquabreeders and culturists should collect founder stocks from as wide a distribution as possible within the species range in their

efforts at domestication. This is to ensure that domestication, at least in its initial stages, be based on the broadest genetic base as possible. These stocks should also be subjected to a wide variety of genetic analyses to determine that the founder group represents, more or less, the same species and that the individuals are chromosomally (genetically) compatible.

3. Research on artificial propagation techniques, including those for in vitro fertilization and development, should be given high priority in order to aid programmes of genetic preservation during the initial stages of domestication. The use of specialized breeding techniques, such as gynogenesis, are being successfully applied in domesticating some species of fish and should be further developed as a part of efforts to bring new species or genetic groups under domestication.
4. The efforts of aquabreeders, culturists and researchers should be directed not only at preserving and maintaining the present domesticated stocks of strains in culture units but also at preserving and maintaining the wild relatives of these genetic groups as valuable reservoirs of genetic variation. Further, it is important to document the process of domestication including the source and history of wild stocks.
5. Research aquabreeders, culturists and fishery managers should be cognizant of the limitations of current methods for estimating genetic variation especially insofar as these techniques may be used for determining a priori which species, geographic population or groups are to be selected for domestication or for restocking natural waters.
6. Hatcheries carrying out introduction or restocking programmes in natural areas should be cognizant of the fact that there may be genetic changes of natural ecosystems which result from using inbred populations. Such stocks are suitable only for "put and take" fisheries, and then only if they are not likely to breed with native stocks. They should also recognize the possible genetic and ecological consequences on natural fish stocks and fisheries of the use of genetically biased brood stocks collected from one area to generate seed for restocking another area.
7. Aquabreeders, research aquaculturists and managers of natural fisheries should direct some of their efforts toward developing methodologies,

proce dures and genetic breeding systems for the generation of suitable genetic groups for stocking deteriorated natural fisheries (especially inland ones). This effort should also involve methods to assess the productivity and survival of properties of the released group.

8. Morphological, meristic and electrophoretic techniques should be used where appropriate to survey commercially exploited fish species to determine specific indicators of populations within each species. When populations can be identified by unique phenotypes or gene frequencies, they should be monitored to prevent overfishing of specific groups thereby reducing the probability of eliminating the genetic resources of unique populations.

9. Previous groups have recommended that there be established an international system for designating strains and stocks of fishes used in aquaculture and hatcheries. Whereas some may feel it is still impractical to do this, we believe that breeders should begin this process for certain fish, for example, common carp, rainbow trout, Tilapia nilotica, and that the discussions could begin at a future international fish breeders or aquabreeders meeting. The FAO Plant Genetic Resources office could provide background based on their broad experience in the preservation of geneticaly and geographically defined natural races.

Recommendations Primarily for the Conservation Community

1. There should be increased effort directed at identifying populations of exploited or endangered species, at solving the major taxonomic problems existing in some parts of the world (especially South America and Africa), and at understanding the role of fish species in the ecosystems in these areas.

2. A series of international meetings should be planned over the next few years in order to define where and how marine and aquatic parts (Biosphere Reserves) should be established. Each meeting or working group should bring together regional and scientific expertise on a specific habitat. These should include, for example, mangrove, coral reef, brackish water, large tropical river and large tropical lake habitats. To facilitate the development of marine and aquatic parts, a working group should be established to discuss and determine sizes and geographic form of natural aquatic genetic reserves taking into account ecological, demographical and genetic criteria.

3. Assistance should be given to fish hobbyists and aquarium managers in maintaining stocks of rare or endangered species, as is already being done in zoos of birds and mammals, giving full attention to the need to maintain adequately large populations of each species.
4. Consideration should be given to developing national or international monitoring schemes for fish, such as the wild trout and char watch at present being organized.
5. There is a general need for identifying important geographic areas, species and their distributions in order to provide background for making decisions about where, what and when to begin any preservation programmes for potentially threatened fishes. Conservationists should cooperate with such agencies as the IUCN in cataloguing threatened fish genetic resources.

Recommendations Primarily for the Scientific Community

1. Physiological "races" have been recognized for many years, yet we do not know whether physiological differences among fish populations are representative of genetic differences. Physiological variation is influenced by environmental variables making the genetic bases of these variables difficult to elucidate. Research on this problem is urgently needed.
2. Monitoring the genetic variability of both natural and cultured fish populations would be advantageous for a number of reasons. However, it is not known if the different methods of ascertaining genetic variability are representative of genomic variance per se. Therefore, research in this area is also highly recommended. Correlations between electrophoretic, meristic and morphological variation should be examined in numerous fish species to determine which methods provide unbiased estimates of genetic variation within species. These studies should be supported, when possible, with artificially induced changes in genetic variation via inbreeding studies.
3. Genetic (electrophoretic) monitoring of several fish species and populations during the entire course of their exploitation is deemed desirable for several reasons: e.g. (1) the possibility exists that similar patterns of change will be detected in different exploited populations; (2) biochemical correlates of changing age or size class structure might be found and used as indicators of overexploitation.

4. Recent advances in genetic engineering techniques suggest that transfers of genetic information between unrelated fish species will be possible within the foreseeable future. Research in this area is to be encouraged because such developments will broaden the genetic resource base for fisheries.
5. Research should be accelerated on the cryopreservation of fish sperm, ova and embryos.
6. Extensive ecological and systematic (taxonomic) surveys are required in tropical regions where large fractions of ecosystems are poorly understood and a large fraction of species are undescribed.
7. Research is needed on the controllability of sterile hybrids in relation to their impact on the ecosystem. It is particularly important to assess the possible genetic "leakiness" of "sterile" hybrid stocks and their possible impact on wild populations.

References

Falconer, D.S., 1960. *Introduction to quantitative genetics*. London, Oliver and Boyd

Helsop-Harrison, J., 1974. Genetic resources conservation: the end and the means. *J.R.Soc.Arts,* Feb. issue:157 p.

Maitland, P.S., 1979. The status and conservation of rare freshwater fishes in the British Isles. *Proc.Brit. Freshwat.Fish Conf.*, 1:237–48

Moav, R., T. Brody and G. Hulata, 1978. Genetic improvement of wild fish populations. Science, *Wash.*, 201(2):1090–4

Villaluz, D.K., 1966. *The Lake Lanao fisheries and their conservation.* Manila, Bureau of Printing, 53 p.

8

Safety Assessment of Genetically Modified Fish

A safety assessment of foods derived from GM animals must be grounded in an understanding of transgenesis methods, anticipated applications and possible outcomes from transgene integration and expression.

TECHNIQUES AND APPLICATIONS

Techniques

A number of techniques can be utilized for transferring genes into animals. They differ in their suitability for different classes of animals, in their efficiency of transformation, and in their implications for risk assessment.

Utilization of the gene transfer approach depends upon knowledge of a gene encoding a product conferring a trait of interest. The gene to be transferred is incorporated into an expression vector that also contains genetic elements to control its expression. The use of different types of expression vectors poses methodological advantages for different classes of animals, and also affects the likelihood of subsequent genetic or immunological hazards being realized. Biotechnologists may purposely transfer into a host a:

— *Fusion gene* - a gene encoding a product of interest with an element that will regulate its expression in the host.

— *Transposon* - a DNA element capable of excising itself from one location in the genome and inserting itself into another location, which has been modified to contain the fusion gene.

— *Retrovirus* - a virus that can integrate itself into the genome and become expressed through the host cell's replicative processes, and that has been modified to contain the fusion gene.

Many expression vectors contain marker genes. Some marker genes are simply reporters for successful gene transfer, while others encode gene products so that transgenic individuals can be selected for, e.g. by application of antibiotics.

Common methods for introducing an expression vector into the host include:

— *Microinjection* - direct injection of the expression vector into fertilized eggs or host cells using a fine glass needle.

— *Electroporation* - introduction of the expression vector into fertilized eggs or host cells by application of pulses of electricity to induce transient pores in the membrane of host cells.

— *Particle bombardment* - coating the expression vector on to gold particles, and introduction in host cells by bombardment with the particles.

— *Cell transformation, followed by cloning* - since it is more straightforward to add or knock out genes for cultured cells than for fertilized eggs, nuclei from successfully transformed cells can be transferred into enucleated eggs and implanted into surrogate dams to generate somatic cell cloned animals, which are also transgenic.

— *Transformation of gametes* - genes may be introduced into oocytes or spermatocytes, and the transformed gametes used for fertilization, generating a whole animal.

Application of any one of these methods will result in the successful transformation of a small percentage of the animals so produced. Transgenic individuals can then be identified and bred to develop a transgenic line.

Applications and their Potential Benefits

Transgenic animals expressing one introduced gene have been or might be developed for a variety of applications, posing a range of possible benefits to food production or human health . Such animals are at various stages of development. Early applications for approval of transgenic animals for food production involve several species of fishes expressing introduced growth hormone genes.

Production of transgenic agricultural mammals is challenging and expensive, especially because of their low reproductive rate and internal fertilization and development. Many transgenic founder individuals are mosaic for the transgene, i.e. they have it in some, but not all, of their cells. For these reasons, development of transgenic agricultural mammals has lagged. However, nuclei from transformed cultured cells or transformed cells from a mosaic animal can be used as donor material for somatic cell nuclear transfer-mediated cloning, producing individuals that are transgenic in all cells. This approach might eventually be used to produce transgenic lines for food production, as is already being applied for development of pigs intended for xenotransplantation, where several transgenes will have to be expressed and several host genes knocked out.

Food-related Hazards Associated with Producing GM Animals

The discussion below addresses hazards associated with transgenesis methods and the environmental release of GM animals that have a bearing on food safety. The possible hazards of transgenesis discussed below must be placed into perspective by considering their relative likelihood and the degree of harm they pose. We note that these hazards are not unique to transgenesis.

Transgenics

Introduction of a transgene into an animal is not a precisely controlled process, and can result in a variety of outcomes regarding integration, expression and stability of the transgene in the host.

The desired outcome generally is stable integration of a single copy of the transgene into a single location in the genome, and not in a functional gene or a regulatory element. However, other outcomes are frequently observed, including integration of multiple copies of the transgene at one locus or insertion of the transgene at multiple locations in the genome. Insertion of the transgene into a host gene may turn the host gene off, sometimes affecting the viability or health of the host. Insertion of a transgene sometimes can affect expression of another gene(s). A transgene may become rearranged before integration, thereby becoming non-functional. During the process of transgenesis, undesired DNA sequences may become inserted into the genome, such as marker genes or selectable markers from the expression vector or contaminating bacterial DNA left over from vector production. Hazards stemming from insertional events or genetic instability can be identified by screening and managed by culling individuals that have undesired events during the course of development of the transgenic line.

Expression of the transgene ideally should have no undesired effects on the expression of other host genes or health of the host. Other outcomes, however, have been observed. The transgene can be silenced by methylation or through other mechanisms. Because expression of the transgene often is controlled by novel regulatory elements outside the host's normal homeostatic feedback mechanisms, expression of the transgene can have pleiotropic effects, that is, effects upon multiple traits of the host. Notable pleiotropies have been observed among animals expressing introduced growth hormone genes, and have included pigs, sheep and fish exhibiting a range of morphological or metabolic abnormalities. Other pleiotropies, such as increased carcass yield, may be positive. Ectopic expression of the transgene may occur in tissues, sexes or life stages where it is not expected, and may affect the health of the host and the safety of its food products. Hazards stemming from transgene expression can be identified by screening and managed by culling individuals with undesired expression phenotypes during the course of development of the transgenic line.

The use of viral and transposon vectors poses the hazard that the transgene might subsequently move within the genome. Work with *Drosophila* suggests that transposons may have a greater probability of movement following crossing into a new background strain. Even though the vectors were engineered to lack all the DNA sequences needed to be packaged into virions or to transpose, there is a theoretical possibility that they could recombine with other DNA sequences in the genome, such as endogenous transposable elements, or with exogenous viruses or transposons, thereby gaining infectivity or mobility. Development and use of well-designed vectors will reduce the likelihood of these hazards.

The development of pigs for xenotransplantation involves knocking out expression of molecules that elicit immune response in humans and adding molecules that make the surface of pig cells more like that of human cells. This raises the possibility that pigs might become more susceptible to human viruses. This could provide an alternative host for spread of human disease, and could also give rise to a new evolutionary pathway for adaptation of pig viruses to humans. This hazard could theoretically be minimized by using pig breeds lacking known endoviruses for development of xenotransplantation lines and by maintaining such lines in strict quarantine.

Cloning

Cloning may be used to propagate GM animals and raises its own issues.

Somatic cell nuclear transfer-mediated cloning requires reprogramming of the genome from a differentiated cell to allow it to drive embryogenesis. This results in some degree of altered gene expression, especially early in the life of the cloned individual. The effects of altered gene expression and of reproductive manipulations needed for the cloning process may result in high rates of prenatal and postnatal mortality and of morphological or physiological abnormalities in cloned individuals, which may in turn affect animal health and food safety. Observation of the limited numbers of offspring of cloned animals produced to date suggests that they may be phenotypically normal.

Environmental considerations that can affect food safety

Different GM animals pose different potential environmental benefits and risks. This discussion does not address all environmental issues but, rather, focuses on the connection between environmental entry of GM animals and food safety. The potential spread of GM animals or their transgenes in the environment is an environmental hazard that provides a route for entering into the human food supply.

The potential entry of GM animals into the food supply via the environment will vary owing to different predispositions of the animals to enter the environment and spread, differences in the farming system's ability to reduce animal escape, and differences in whether humans hunt or fish for the same species. Some farmed animals are often transported and sold alive, posing additional routes for accidental entry into the environment. Escaped GM fish and shellfish, or their descendants, could be harvested without being detected and subsequently eaten by people. The current status of development of GM animals suggests that food safety managers might be faced with this issue first for GM fish and shellfish, and somewhat later for some kinds of GM poultry such as ducks and quail.

Key species or taxa of GM animals can be ranked in terms of their ability to become feral, likelihood of escape from captivity, mobility and historical reports of ecological community disruption Such a ranking from high to low for North America. would consist of insects, shellfish, fish, mice-rats, cat, pig, goat, horse, rabbit, dog, chicken, sheep and cattle. On a regional scale, the relevance of this ranking to food safety will change depending upon whether these animals are widely eaten by humans. Furthermore, the rankings will vary among regions owing to different environmental conditions, but the same risk factors apply.

Assessment of the potential environmental spread of GM animals or their transgenes should be done case-by-case for each combination of integration event (i.e. transgenic line) and local environmental conditions. The assessment should compare the GM animal with its conventional counterpart, i.e. unmodified animals derived from the same genetic source. The assessment should estimate the probability of movement of the GM animal or its transgene(s) into the environment, given the estimated rate of escape. This involves assessing whether:

— the GM animal, compared with the conventional counterpart, has a lower, equal or higher potential for gene flow to any wild or feral relatives found in the receiving ecosystem. Recent research suggests that the transgene could be purged within a few generations or could spread through the natural population and possibly affect its abundance. For potential entry into the food supply, purging would be a safer outcome. However, these animals may still enter the food supply because the purging process is likely to take one or more generations;

— the GM animal, compared with the conventional counterpart, has a lower, equal or higher potential to invade and establish itself as an alien species, particularly when the receiving ecosystem lacks wild or feral relatives.

Status of methods for estimating potential environmental entry

The best methods for reliably characterizing potential environmental entry have not yet been standardized. The net-fitness methodology provides a systematic and comprehensive approach based on contemporary evolutionary and population biology. It involves a two-step process: (1) measuring fitness-component traits covering the entire life cycle for GM animals, their conventional counterparts or wild relatives, and crosses between the two; and (2) entering the fitness data from step 1 into a simulation model that predicts the fate of the transgene across multiple generations. There is a need to validate this method's predictions; initial experiments are under way towards this end. For example, a transgenic fish project is testing the validity of the net-fitness model's predictions. There is also a need to add stochasticity, elaborate additional features and improve the user-friendliness of the simulation model. The data needed to apply this methodology have yet to be obtained for most GM animals; efforts to gather such data have recently begun for a few cases of GM fishes.

Confinement of GM animals

If the combined conclusions from the environmental hazard characterization (discussed above) and the food safety assessment are that GM animals or their transgenes will spread in the environment to a degree that poses a risk to the human food supply, then risk managers should consider the need to apply confinement measures to prevent or reduce escape of GM animals or their viable gametes into the environment. The primary focus of these measures should be to ensure that release does not occur. If this cannot be assured, then it can be complemented by the use of methods to ensure that any escaped individuals cannot reproduce.

Biological, mechanical and physical/chemical confinement measures are available for fish and shellfish produced in different aquaculture systems. Biological confinement measures typically involve disrupting the animal's ability to reproduce, such as sterilizing fish and shellfish by induction of triploidy, i.e. resulting in individuals with three sets of chromosomes rather than the normal two sets. Mechanical confinement involves application of some kind of implement to prevent or reduce escape of animals from the aquaculture system (e.g. screens in effluent pipes of land-based fish tanks or ponds) and physical confinement involves making an aqueous escape pathway lethal by changing a physical attribute of the water (e.g. heat effluent water to lethal temperature, and then cool down before discharge). Other confinement systems could be developed for GM terrestrial animals.

In many cases, there is a need for multiple confinement measures because no single measure is fully effective. For instance, all-female triploid populations of salmon can be used to ensure that any individuals that might escape physical confinement are unable to reproduce in the wild. There is also a need for robust verification of confinement measures.

There is a need to develop and validate better methods for reliably inducing reproductive sterility of GM animals, in particular GM fish and shellfish. Improved methods could include repeatable protocols for induction of triploidy (applicable to animals other than birds and mammals) and new methods for inducing sterility via chemical treatments or gene transfer, for instance, by inserting anti-sense genes to disrupt key steps in the endocrine pathway controlling reproductive development.

Monitoring for environmental entry and spread

In future, specific GM animals may gain approval for widespread production, either with or without approval to enter them in the human food supply. In such situations, it will be important to consider whether or not to apply post-market monitoring for unexpected environmental spread of the GM animals and their transgenes that pose food safety hazards. Methods for detection of such GM animals and their transgenes in the environment are likely to involve the application of two well-established bodies of scientific methodologies: (1) diagnostic, DNA-based markers and (2) sampling protocols that are adequate (in terms of statistical power) and cost-effective. However, there is a need to develop fully appropriate protocols for the application of these methods to post-market detection of environmental spread of GM animals and their transgenes. Monitoring can also be helpful to assure confinement of GM animals during research and development.

Future Perspectives

Realizing the full range of potential benefits from use of GM animals will depend on advances in technical aspects of their production. For instance, innovative molecular methods will be needed in order to address current limitations with respect to low frequency and randomness of integration, gene silencing and epistatic and pleiotropic effects of transgenes. These innovations may include using improved expression vectors to target transgenes to specific places in the host genome or incorporating transgenes on to bacterial or yeast artificial chromosomes and their introduction into the host. Greater experience with anti-sense gene expression and homologous recombination-based gene knockout techniques will allow the turning off of targeted genes. Other advances, especially development of embryonic stem cells or primordial germ cells from additional species, will facilitate a greater range of genetic modifications of those animal species. Identification and the transfer of genes will promote beneficial food uses, and also necessitate their assessment with respect to food safety and environmental impacts (e.g. insects with food safety implications, such as honey bees). Transfer of new genes and advances in gene transfer and cloning techniques will facilitate developments contributing to human health by means of new animal models of human disease, expression of pharmaceutical proteins and development of genetic lines for xenotransplantation. Although many potential benefits from GM animals can be anticipated, these will present more challenging conditions for risk assessment, risk management and risk communication.

Approaches to the Safety Assessment of GM Animals

The food safety assessment of GM animals and derived products can largely be performed along the lines that have already been established for the evaluation of GM plants and derived products for the consumer. This means that the initial step of the food safety assessment will be a comparative safety assessment of the GM animal with its appropriate comparator, including a food intake assessment, followed by a full risk characterization.

General Principles

Comparative Safety Assessment

Because of their history of safe use, it is generally acknowledged that traditional food products may serve as a baseline for comparison for the assessment of the safety of GM foods. This is generally referred to as the concept of substantial equivalence. The basic idea is that novel genetically modified organism (GMO)-derived food products should be at least as safe as the traditional products that they may replace in the diet.

Substantial equivalence is not a safety assessment in itself; rather it represents the starting-point which is used to structure the safety assessment of a GM food relative to its conventional counterpart. Since its inception, the concept has evolved. More recently, it has been suggested to replace it with the broader Comparative Safety Assessment (CSA) concept.

The CSA is basically a two-tiered approach. The initial step comprises a thorough comparison with the closely related conventional counterpart to identify any differences that may have safety implications for the consumer. This comparison includes both phenotypic characteristics as well as a compositional analysis. The phenotypic analysis should also include comparative health parameters. The compositional analysis will focus on key substances in the animal products under scrutiny and will be subject to changes according to the latest scientific state of the art.

The second step of the CSA comprises the toxicological and nutritional evaluation of the identified differences between the GM animal and its conventional counterpart. As a result of this second step additional testing may be required and can result in an iterative process in order to obtain all relevant information for the final risk characterization.

In the initial step the information that should be considered includes:

— the transformation process of the genetic modification, including the sequence of the inserted material before and after the transformation event;
— the copy number and site(s) of insertion;
— sequence analysis of the site(s) of insertion, i.e. flanking regions;
— stability of the integration (multiple generations);
— the safety of any newly expressed proteins, including assessment of allergenicity;
— occurrence and implications of unintended effects;
— the role of the new GM animal food in the diet; and
— the potential influence of processing or spoilage on the new GM food product.

More precise criteria for the molecular characterization are dealt with in the Codex guidelines (Codex Alimentarius Commission, 2003) and are currently being discussed for further elaboration within the OECD.

In terms of classical risk assessment, the CSA approach means that for complex food products, including GM animal-derived products, it will in general be necessary to do both hazard identification and characterization as well as the food intake assessment in order to be able fully to characterize the risk.

Hazard Identification and Characterization

Hazard identification and characterization are typically the first steps in any risk assessment. Any differences found as a result of the CSA serve as comparable to the hazard identification and hazard characterization steps in a traditional risk assessment paradigm. However, for complex GMO-derived foods, the hazard identification and characterization steps will not be as readily completed as in the case of well-characterized single chemical compounds because of the variety and magnitude of unintended effects that may occur when testing complex food products.

Molecular characterization

An extensive molecular characterization of the inserted genetic material construct will generally be required, both before and after the insertional event. The molecular characterization should furthermore comprise an analysis of the copy number and a sequence analysis of the flanking regions of the place of insertion in order to identify any unintended effects.

Safety of the gene product

The safety of the gene product must be assessed on a case-by-case basis. Depending on the knowledge of the expressed product, the assessment may range from a limited evaluation process of the available data on the protein, such as amino acid sequence and expression rates in different tissues, to, in the case of less well-documented proteins, extensive toxicity testing including animal studies. In theory, production of GM animals may lead to the introduction of many new proteins without a history of safe use into the human diet. The assessment of the novel proteins should be based on current knowledge of toxic substances, including a search for sequence homology with known toxins, and the function of the novel protein. In the case of unknown proteins, a full classic toxicological safety assessment procedure will form part of the evaluation.

In this regard, a distinction should be made between GM animals developed for food purposes and GM animals developed for pharmaceutical, xenotransplantation or industrial purposes. While all GM animals and their derived products in the latter category will not be intended for food purposes, such animals may enter the food supply either directly or indirectly and therefore their hazards should be assessed.

So far, the number of different genes that is used for the production of GM food animals is still rather limited when compared with plants, but this situation may change with the progress of genome sequencing programmes that are likely to provide a wealth of data on important animal physiological pathways.

Allergenicity

In the case of newly expressed proteins in the GM animal, the allergenic potential of the protein will need to be assessed. For the production of specific well-characterized proteins by the GM animal, it needs to be established whether post-translational modifications are comparable to the same substances being produced by more traditional sources in order to assess potential altered toxicological or allergenic properties of the newly synthesized proteins.

It has been recognized that there is no single parameter that can predict the allergenic potential of a substance. Recently, a strategy to assess allergenicity of biotechnology products has been formulated, which relies on the following parameters: source of the gene, sequence homology, serum testing of patients known to be allergenic to the source organism or to sources distantly related, pepsin resistance, the prevalence of the trait and assessment using animal models.

The strategies and methodologies for the allergenicity testing in GM animals will not differ fundamentally from those currently in use for the assessment of GM plants. It was recognized that animal models for allergenicity testing, even those that are not yet validated, may be of value to identify potential allergens. It is recommended that additional efforts should be directed to the further development and validation of these models.

Gene transfer

The DNA construct used to change the genetic make-up of the animal should be considered within an assessment, especially if the gene or its promoter is derived from a viral source, since horizontal transfer or recombination may occur. Additionally, bacterial host-derived materials may include additional sequence fragments unrelated to the target gene. Inadvertent introduction of such sequences into the germline of a GM animal not only has the potential for creating unintended genetic damage but can also contribute by recombination to the generation of novel infectious viruses. A well-known example is the generation of a replication-competent murine leukaemia virus (MLV) during the growth of a vector containing a globin gene.

There is also potential for horizontal transfer of the gene construct: food-ingested foreign DNA may not be completely degraded in the gastrointestinal tract of mice and pigs. For the food safety assessment, it is prudent to assume that DNA fragments may survive the human gastrointestinal tract and be absorbed by either the gut microflora or somatic cells lining the intestinal tract.

Assessment of the safety of the genetic construct should include marker genes. Commonly used marker genes are genes that code for antibiotic resistance. Risk assessment of these selectable genes should focus on gene transfer to micro-organisms residing in the gastro-intestinal tract of humans or animals. However, as the potential of this gene transfer cannot be completely ruled out, the safety assessment should also consider information on the role of the antibiotic in human and veterinary medical uses.

Unintended effects

Potential unintended effects represent a significant concern with GM animals and these effects highlight the difficulty of establishing generic considerations instead of case-by-case considerations. Unintended effects can be divided into insertional effects, related to the place of insertion of the transgenic fragment, and secondary effects, related to the nature of the expression products of the introduced genes. The major approach to detect any unintended side effects in the GM animal is phenotypical, including a compositional analysis to compare the new food organism with the conventional counterpart.

In general, the compositional analysis should be performed on the basis of validated scientific methods. Strategies for the compositional analysis of food products derived from GM animals will not differ fundamentally from those of plants, where key substances are identified and analysed per species. Furthermore, in order to be able to interpret the data from the compositional analysis of individual animal products adequately, insight into the natural variation in the relevant macro- and micronutrients and antinutrients, if present, will be required.

A special case is hemizygous GM animals that are intended for marketing. If the trait is inherited in a recessive way, so that it is expressed only in a recessive homozygote, it may be prudent to assess both the hemizygote as well as the homozygous animal for potential deleterious effects of the specific trait.

In future, compositional analysis may also be based on unbiased profiling of the GM food product and the conventional counterpart. Techniques for the profiling approach are now under development and can be divided into three subsections: genomics, proteomics and metabolomics to screen for differences in the GM animals in relation to the gene transcription products, proteins and metabolites, respectively. At the moment, however, none of these techniques is yet validated and ready for routine use in risk assessment.

Food Intake Assessment

The goal of a food intake assessment is to assess the amount of food or food ingredient an individual or population group may consume. No exact criteria have been formulated so far for the factors that need to be considered in a pre-market intake assessment of a complex novel food product. Some food intake paradigms make assumptions based on per capita production while others use per capita distribution. An intake assessment may also consider the cooking and food preparation process used. Some governments have instituted tracking of animal-derived food and from this data set post-market consumption data may be determined. Food intake assessments will also include an estimate of the extent to which current food products will be replaced by the GM animal-derived novel food product. Thus, the accuracy of the intake assessment for GM animal-derived foods is dependent upon the available data on consumption patterns of consumer groups of interest and the validity of the underlying parameters. Specific consumer groups may refer to different age groups, but also to more vulnerable groups such as pregnant or lactating women or specific patient groups.

Food intake assessment will be based not only on available consumption data, but also on our knowledge of the bioavailability in the gastro-intestinal tract of specific food components under investigation. Probabilistic mathematical models for integrating food consumption and distribution may in specific cases be used in a comparative approach to estimate future intakes more precisely.

Integrated Toxicological Evaluation

Following the phase of hazard identification, hazard characterization and food intake assessment, an integrated toxicological evaluation will combine all the information in relation to the food safety of the complex GM animal-derived food product. This integrated toxicological evaluation needs to identify food safety issues that may require additional investigation, including traditional toxicity testing.

In general, it will not be possible to test complex animal products by classical toxicological animal studies in the way they are routinely used to test single compounds. Classical studies measuring physiological responses relative to dose are complicated if the laboratory animal is receiving doses of the GM animal's edible tissue. If the genetic modification would result in the expression of novel proteins or if the compositional analysis revealed an alteration in an endogenous protein product or metabolite, the traditional toxicological approach would require the concentration of the product to be elevated in the laboratory animal's diet to the extent that the diet will often become unbalanced. This might result in toxicological observations that are unrelated to the product under investigation. The limitations of standard toxicity testing applied to whole foods have also been discussed.

On occasions where the genetic modification results in an increase in a specific (exogenous) protein, for instance directly derived from the gene construct, traditional testing would still be valid to assess that protein. Alternatively, there may be instances wherein endogenous protein levels in the GM food are increased well above the physiological level in the given animal species and it might be prudent in specific cases to (also) test this elevated protein in animal studies.

Integrated Nutritional Evaluation

In order to identify nutritional issues that need further investigation, an integrated nutritional evaluation will need to combine all the information related to the nutritional aspects of the complex GM animal-derived food product. This has to be done in addition to the integrated toxicological evaluation.

The nutritional analysis should focus on the potential replacement of nutritionally important food products by the novel GM animal-derived food products with possibly altered characteristics. The information for the nutritional analysis will largely be derived from the initial CSA, including the compositional analysis (especially macro-, micro- and antinutrients) and the estimated consumption rates. Detected alterations in the GM animal-derived food products compared with the traditional counterpart will be assessed by evaluating the significance of the compositional differences for the consumer in general and also, in specific cases, for specific consumer groups. Nutritional aspects of GMO-derived foods may become of increasing significance when the number of compositionally altered food products on the market increases. Therefore, the nutritional assessment of GM animal-derived food products is dependent on current consumption data of animal-derived food products in distinctive consumer groups and with respect to geographical and demographical differences. Special consumer groups perhaps worthy of consideration include children, pregnant or lactating women, elderly persons and the immuno-compromised.

Micronutrients are vitamins and minerals that are essential for normal physiology and biochemical functioning. Both deficiency and excess of a micronutrient can cause health problems which emphasizes the importance of this class of compounds. Macronutrients include dietary lipids, proteins and carbohydrates and these classes of compounds are present in the food and diet in substantial quantities. Assessment of the replacement factor of important animal-derived sources of micro- and macronutrients by GM animal products in the event of altered composition with relation to these nutrients is therefore of major importance. Bioavailability of the important micro- and macronutrients from GM animal-derived tissues is also of significant importance in this respect.

Risk Characterization

Risk characterization is the final step of the risk assessment process and involves integrating the outcomes from the full toxicological and nutritional evaluations in order to reach an overall conclusion about the safety of the food.

The baseline for the safety of novel food products derived from GMOs, including GM animals, in all cases will have to be the assessment that the novel GM animal-derived food products is at least as safe as its conventional counterpart. If any questions remain after the initial CSA with respect to the safety of the GM animal-derived food products, additional tests may be required, including animal studies with the whole product or selected tissues/ extracts. If, after a full safety assessment, the safety standard, i.e. as safe as its conventional counterpart, cannot be satisfied, the GM animal-derived product should not be approved for marketing. This risk characterization should be established on a case-by-case basis for food products derived from GM food animals.

Post-market Surveillance

In general, potential safety issues should be addressed adequately through a rigorous pre-market assessment, as the feasibility of post-market studies is very limited at present.

However, post-market surveillance may be an appropriate risk management measure in specific circumstances. Its need and utility should be considered, on a case-by-case basis, during risk assessment and its practicability should be considered during risk management.

The use of post-market surveillance as an instrument to gain information on the potential long-term or unexpected adverse and beneficial effects of food, either GM animal-derived or traditional, should be further explored. Post-market surveillance could be useful in certain instances where clear-cut questions require, for instance, a better estimate of food intake and/ or nutritional consequence of a GM animal-derived food.

For GM animal-derived medicinal substances, existing pharmacovigilance schemes will apply to monitor any unforeseen unintended side-effects of the isolated medicinal substances. The same would apply in a veterinary sense with respect to the GM animal itself when modified with respect to the production of hormonal or disease-prevention substances: pharmacovigilance schemes could help to detect unintended side-effects of the introduced expression product to the GM food animal that were not detected in the pre-market phase. To this end, the GM animals should then be included in such pharmacovigilance schemes on the basis of "internal" administration of the specific veterinary substance.

It is important to note that product tracing and related control systems may be less straightforward in the case of chimeric organisms as different parts of the food animal will have different genetic constitutions and this may severely complicate analytical control of product tracing systems.

Depending on the questions and risk management needs underlying the establishment of post-market surveillance systems the product information conveyed to the consumer may, however, require adjustment. In order to enable consumers to relate potential adverse, e.g. allergenic, effects to a GM animal-derived food product, it may be necessary not only to label the product as GM animal-derived, but also to provide information on the specific GM animal source, for instance by including on the label the unique identifier code specific for a single integration event.

For the establishment of adequate product tracing systems and in order to distinguish safety evaluated products from unevaluated ones, it will be necessary to have the information on the sequence of the insert and on the flanking regions as well as reference materials of the transgenic animal and its conventional counterpart. It is recommended to use a relatively small size sequence for detection which is expected to be detectable in all kinds of products, including processed products.

Specific Food Safety Issues

This section deals with specific issues that are frequently raised with regard to the safety of genetically modified foods. These issues include phenotypic analysis, sampling, compositional analysis and post-market surveillance. The discussion which follows provides an evaluation of existing knowledge about these topics and elaborates scientific approaches that may be used to assess possible health risks.

Phenotypic Analysis

The selection process of the initial founders will be very limited compared with the plant breeding situation where thousands of GM calluses are screened for incorporation of the transgenic fragment and subsequently monitored for their phenotypic characteristics. This means that the information on the variation range between animals with the same genetic modification will be rather limited and that detected differences between individual animals will be difficult to interpret.

Phenotypic characteristics

Phenotypic analysis will relate to compositional analysis, but also to general performance parameters (such as growth rate, feed conversion efficiency, reproduction, clinical parameters), disease resistance and, depending on the animal and the genetic modification, the colonization and shedding of pathogens that may have food safety implications.

In specific cases, phenotypic analysis may also be advisable after processing or, for fish, during the various stages of spoilage. For example, adverse biogenic amines can be formed during spoilage in salmon, tuna, herring and other fish species. Similarly, formaldehyde may be formed in spoiled shrimp, cod, hake and many other species.

Sampling

For individual animal species the number of GM animals and conventional counterparts required for a compositional analysis needs to be determined in order to result in statistically reliable results, i.e. perform a power analysis. Furthermore it will need to be decided which edible tissues and products should be analysed in the different animal species.

It is clear that, in the case of GM fish and other aquatic species, more individuals are likely to be available for a compositional analysis compared with, for instance, large farm animals. In order to obtain statistically meaningful results in both cases it will be necessary:

— to have information on the natural variation in the different tissues;
— to apply standardized experimental conditions to the GM animals and conventional counterparts; and
— to have standardized conditions for harvesting tissues and other animal products, for instance, with respect to developmental stage, age or market weight.

An example is the GM fish that has incorporated a growth hormone gene into its genome. In this case, sampling for the compositional analysis may be based on market weight rather than age.

Compositional analysis

For GM animal-derived food products the same basic approach for the compositional analysis should apply as for plants. Similar to the GM plants, the key constituents of the tissue would have to be established. This will include key nutrients as well as those compounds that may have adverse effects on human health, such as thiaminase in fish and wax esters in butterfish. The list of key constituents per tissue will have to be flexible and may have to be adjusted based on the current state of the art in an ongoing process.

It is important to generate background data on the natural variation for the individual constituents in different tissues. Existing databases on the composition of animal food products need to be evaluated as to whether their data are of sufficient quality to be of value in a comparative compositional analysis.

Post-market Surveillance

Post-market surveillance requires food intake data and may require the establishment of adequate product tracing systems. In this respect the food animal sector has advantages over the crop plant sector where basic product tracing systems are still virtually lacking. In the animal production sector, such systems are already well established for some animal food production chains in some countries and many other initiatives are ongoing in this field. It will, however, require further elaboration and adjustment before these systems can be used for the purpose of post-market surveillance.

Future Developments

As a result of ongoing genome programmes, the potential of genetic modification in animal production will increase. The variety of genes that can be used will increase and it may also become more feasible to transfer entire metabolic pathways. It may be anticipated that the number of GM animal-derived products with improved health characteristics may increase, for instance pigs with a nutritionally improved meat composition or shrimp with reduced allergenic potential.

Based on our increased knowledge of the physiology of food animals, the number of unpredictable secondary effects of the genetic modification that cannot be analysed in a targeted analysis may decrease, theoretically reducing the number of unexpected effects of a transformation event.

Further development and validation of profiling methodologies in the field of genomics, transcriptomics, proteomics and metabolomics may provide additional insight into the unintended effects of a transformation event.

Improved product tracing and information transfer systems may make the application of post-market surveillance systems more feasible to assess the long-term effects of complex food products, including food products derived from GM animals.

International Regulatory Instruments

Codex Alimentarius

In July 2003, the Codex Alimentarius Commission adopted the following texts:

Codex Principles for the Risk Analysis of Foods Derived from Modern Biotechnology

The purpose of these principles is to provide an overarching framework for undertaking risk analysis of the safety and nutritional aspects of foods derived from biotechnology. The "tracing of product" is referred in the text as a specific tool to facilitate risk management measures.

The key elements of the principles are:

— there should be a pre-market food safety assessment, on a case-by-case basis, for foods derived from biotechnology. The assessment should be based on sound science, obtained using appropriate methods and analysed using appropriate statistical techniques. The data and information used in this assessment should be of a quality that would withstand scientific peer review;

— the food safety assessment is based on a comparative analysis with a "conventional counterpart" to ensure that the resulting biotechnology food is no less safe than the foods normally consumed by the population;

— risk management measures should be proportional to the risks identified in the safety assessment and may include measures such as labelling, post-market monitoring and product tracing; and

— the definitions used in the Principles for the [Food Safety] Risk Analysis for Foods Derived from Modern Biotechnology are the same as those in the Cartagena Protocol on Biosafety (CPB), so that the Codex texts on food safety and the CPB text on biosafety and environmental protection are mutually compatible and supportive.

Codex Guideline for the Conduct of Food Safety Assessment of Foods Derived from Recombinant-DNA Plants

This guideline, which is based on the above "Principles", describes the methodology for conducting a safety assessment specifically for foods derived from recombinant-DNA plants. The basic approach for the safety assessment is the comparative approach based on the concept of "substantial equivalence", thus focusing on the difference between foods derived from recombinant-DNA plants and their conventional counterpart. It pays special attention to the question of allergenic potential of new genetically modified (GM) plant varieties.

Codex Guideline for the Conduct of Food Safety Assessment of Foods Produced Using Recombinant-DNA Microorganisms

Based on the above "Principles", this guideline describes the methodology for conducting safety assessment specifically for foods derived from recombinant-DNA micro-organisms. The basic approach is similar to the guideline of recombinant-DNA plants, however, elements peculiar to micro-organisms were highlighted.

The key elements of the guidelines are:

— a detailed step-by-step guidance on how to undertake a safety assessment, including the nature of the data to be collected and the elements in the decision-making process that allows food produced using recombinant-DNA micro-organisms to be considered suitable for human consumption;
— they allow comparison of the safety assessments undertaken by different national authorities;
— they allow national authorities that do not wish to do their own (rather expensive) safety assessments to use the safety assessments of other government authorities provided that these assessments are in line with the Codex Guidelines;
— they provide a basis for future food safety assessments that may be undertaken by FAO and WHO, if FAO and WHO decide to undertake case-by-case safety assessments.

Cartagena Protocol on Biosafety

The Cartagena Protocol is a legally binding international instrument that regulates the transboundary movement of living modified organisms (LMOs) resulting from modern biotechnology with the objective of protecting the environment. The backbone of the Protocol is the advance informed agreement (AIA), requiring consent prior to the shipment and introduction of an LMO into the environment of an importing country.

The Protocol establishes a harmonized set of international rules and procedures designed to ensure that countries are provided with the relevant information, through the "Biosafety Clearing-House". This Internet-based information system enables countries to make informed decisions before permitting the import of LMOs. The Protocol also ensures that LMO shipments are accompanied by appropriate identification documentation. The Protocol entered into force on 11 September 2003. It is important to note that, at present, there are no internationally agreed frameworks for considering ethical aspects relating to the use of modern biotechnology.

Ethical Aspects

The production of genetically modified animals raises a variety of legitimate ethical concerns. The need to address these concerns openly and without prejudice is further amplified by sections of a critical public that question some of the achievements of modern biotechnology. Responsible decision-making and policy needs to integrate ethics among the salient factors in the preparatory stages of risk analysis.

Ethics is rooted both in the world religions and in secular philosophies, while a sense of morality and moral value is common for everybody. Ethics comprises both a positive dimension relating to our conceptions of the good life/society, and a negative dimension relating to our judgements of what is morally wrong. For instance, owing to religion-based food practices relating to eating pork, the utilization of genetic material from pigs could present problems.

Environmental Ethics and Animal Welfare

In large parts of mainstream Western ethics, the objects of our moral concerns have been human beings. With the advance of our knowledge, many people have come to realize that there is ample reason to extend the realm of moral concern to animals and perhaps even to ecosystems. This has led to discussions on the moral status of animals, as well as to increased attention being given to animal welfare issues, also related to transgenic animals. For instance, one set of early experiments with growth enhanced salmon showed some individuals with cranial deformities. As a general, but not unexceptional, rule the intended use of GM animals for food production bespeaks the interest of the producer to ensure or improve the health of animals and good animal welfare. Therefore, aspects of animal welfare of transgenic animals need to be evaluated on a case-by-case basis by competent bodies.

Uncertainty

Ethically responsible decision-making demands, *inter alia*, both the utilization of the best available knowledge and an awareness of the relevant uncertainties involved. While it is widely acknowledged that good risk assessment demands a measure of uncertainty, the common instruments to make these uncertainties visible are still limited. However, research on this topic has made significant progress during the last decade, and valuable and useful instruments to represent the relevant uncertainties are now available. This is ethically significant because, in some cases, the uncertainties relate to our state of knowledge (thus often indicating the need of more research) but, in other cases, the uncertainties relate to inherent characteristics of the system under study, e.g. chaos or complexity, or multiple states of equilibrium without linear and deterministic state change. In the latter cases of inherently limited predictability, we have to adopt a responsible scheme for the management of these uncertainties. This point is important both for the use of precaution in risk management and for the provision and presentation of scientific findings as the basis for such management. Precaution does not assume an unrealistic notion of zero-risk. Sometimes the best precautionary action is carefully controlled, monitored and stepwise development. There is a need to address explicitly the uncertainties involved in our assessments, and to adopt schemes for their responsible management.

Transparency and Public Deliberation

The recognition of consumer autonomy and the right of free and informed market choices is an aspect of responsible management. Another aspect relates to the worry in sections of the public that the new genetic technologies may not be used for the "right" or ethically justified ends. The distribution of risks and benefits may be morally problematic, and the scope of benefits may be wanting. At present, even the lack of data on such distributions is a concern. Similar concerns relate to the technological divide and the unbalanced distribution of benefits and risks between developed and developing countries. Often the problem becomes even more acute through the existence of intellectual property rights and patenting that places an advantage on the strongholds of scientific and technological expertise. Equity and fairness issues are thus obviously important. All these considerations point towards the positive dimension of ethics, i.e. a discussion of purpose, benefits and risks. There is a societal need to address these issues upfront, and in the early stages of development. Proactive assessment is indicated, and the need for scientific data to inform such assessments should be described. Risk managers and decision-makers should shoulder this task in collaboration with stakeholders.

The Role of Ethical Principles in Assessments

In relation to human health and medicine there is already a tradition of carrying out practical ethical assessments. Four principles have been established as fundamental in the biomedical field: respect for autonomy, beneficence, non-maleficence and justice. These principles seem to be widely accepted, represent important ethical theories and cover most of the problems appearing in the biomedical field. Within the field of genetically modified animals a similar framework needs to be introduced if ethics is indeed to become an integral part of regulation and guided policy advice. Extensions of the principle approach in biomedicine to other technological and environmental issues have been carried out e.g. in the ethical matrix approach. The basic idea in this framework is to combine the use of a variety of principles with the interest-related perspectives of the various stakeholders and other potentially affected organisms and their environment. A schematic version of one such approach is presented in the next paragraph. The rationale of these frameworks and approaches is to make ethical assessment more transparent and more methodical, and thus amenable to quality assurance.

A Schematic Ethical Assessment

Assuming that we want to assess the ethical aspects of a certain genetic modification of a fish species for food production in a region, following the ethical matrix approach we would first address the issue of who the relevant stakeholders are. We also need to agree on potentially affected organisms and their components of the environment, for example fish and other biota. A proper set of ethical principles then needs to be established, for instance, justice/fairness, dignity/autonomy and welfare considerations as comprising both the elimination of negative welfare and the increase of positive welfare. Once a common understanding of these principles is ensured, it is important to specify the principles for each interest perspective.

References

Beauchamp, T.L. & Childress, J.F. 2001. *Principles of biomedical ethics*. 5th ed. (1st ed. 1979). Oxford, UK, Oxford University Press.

FAO/WHO. 2000. *Safety aspects of genetically modified foods of plant origin*. Report of a Joint FAO/WHO Expert Consultation on Foods Derived from Biotechnology, Geneva, Switzerland. Rome, FAO.

National Research Council (NRC). 2002. *Animal biotechnology: science-based concerns*. Washington, DC, National Academies Press.

Sagar, A., Daemmrich, A. & Ashiya, M. 2000. The tragedy of the commoners: biotechnology and its publics. *Nature Biotechnology*, 18 (January 2000): 2-4.

Scientists' Working Group on Biosafety. 1998. *A manual for assessing ecological and human health effects of genetically engineered organisms*. Edmonds, WA, USA, Edmonds Institute.

Walker, W.E., Harremoës, P., Rotmans, J. *et al.* 2003. Defining uncertainty - a conceptual basis for uncertainty management in model-based decision support. *Integrated Assessment*, 4(1): 5-17.

9

DNA-based Molecular Techniques for the Detection of Fish Pathogens and Diseases

In 1995, global production from aquaculture reached 27.8 million tones and was valued at US$ 42,300 million. Developing countries contributed over 87% of total production, of which 90.1% was from Asia. China contributed 63.4% of total world production. Over the past decade, global aquaculture production has grown at an average annual rate of 9.6% compared to 3.1% for livestock meat and 1.6% for capture fisheries. Between 1984 and 1995, growth in aquaculture production in low-income food deficit countries (LIFDCs) was over five times faster than in developed countries.

Impact of Diseases

Disease outbreaks are recognized as a significant constraint to aquaculture production and trade, affecting both the economic development and socioeconomic revenue of the sector in many countries in the world. According to Chamberlain, disease is a primary limiting factor for shrimp farming today and the risk of disease losses is likely to increase as the shrimp sector continues to grow. Economic loss attributed to outbreaks of disease in developing countries in the Asian region was estimated to be at least US$ 1,400 million in 1990. The cost of lost production in China alone was approximately US$ 1,000 million in 1993. In Thailand, the loss in 1996 due to yellow head virus (YHV) and white spot syndrome virus (WSSV) was estimated to be 40% of total production (70,000 tones) valued at over US$ 500 million. Recent estimates, based on farm level surveys in 16 Asian countries, suggest that disease and environment-related problems have caused annual losses of more than US$ 3,000 million to aquaculture production. Serious financial losses have also been recorded in other regions of the world. In 1993, Ecuador lost 28,000 tones of shrimp production in due to an epizootic of Taura syndrome virus. Salmon farming in many countries also faced serious disease problems that resulted in significant production losses. Various factors have been related to the apparent increased incidence of disease. Environmental factors and poor water quality, sometimes resulting from increased self-pollution due to effluent discharge and pathogen transfer via movements of aquatic organisms appear to be an important underlying cause of such epizootics.

DNA-Based Technologies in Diagnosis and Pathogen Detection

The effective control and treatment of diseases of aquatic animals requires access to diagnostic tests that are rapid, reliable and highly sensitive. In many cases, post-mortem necropsy and histopathology have been the primary methods for the diagnosis of fish and shellfish diseases. However, these methods often lack specificity and many pathogens are difficult to detect when present in low numbers or when there are no clinical signs of disease. Direct culture of pathogens is also widely used. However, these methods are time-consuming and costly, and, for shrimp and other crustaceans, cell lines suitable for virus culture have not been available.

Efforts to overcome these problems have led to the development of immunoassay and DNA-based diagnostic methods including fluorescent antibody tests (FAT), enzyme-linked immunosorbent assays (ELISA), radioimmunoassay (RIA), *in situ* hybridization (ISH), dot blot hybridization DBH) and polymerase chain reaction (PCR) amplification techniques. The use of DNA-based methods derives from the premise that each species of pathogen carries unique DNA or RNA sequences that differentiate it from other organisms. The techniques offer high sensitivity and specificity, and diagnostics kits allowing rapid screening for the presence of pathogen DNA are moving rapidly from development in specialized laboratories to routine application. DNA probes are expected to find increasing use in routine disease monitoring and treatment programs in aquaculture, in field epidemiology and in efforts to prevent the international spread of pathogens (national quarantine and certification programs).

DNA-based methods have been used in diagnosis and for detection of many economically important viral pathogens of cultured finfish and penaeid shrimp. For finfish, tests have been developed for pathogens such as channel catfish virus (CCV), infectious hematopoietic necrosis virus (IHNV), infectious pancreatic necrosis virus (IPNV), viral hemorrhagic septicemia virus (VHSV), viral nervous necrosis virus (VNNV) and *Renibacterium salmoninarum*. PCR has been used in Japan to screen striped jack (*Pseudocaranx dentex*) broodstock for VNNV, permitting selection of PCR-negative spawners as an effective means of preventing vertical transmission of this pathogenic virus to the larval offspring.

DNA-based detection methods for detection of penaeid shrimp viruses are now used routinely in a number of laboratories around the world. These include probes for such diseases as white spot syndrome virus (WSSV), yellow head virus (YHV), infectious hematopoietic and infectious hypodermal and haematopoeitic necrosis virus (IHHNV) and Taura syndrome virus (TSV) which pose the greatest threat to world shrimp culture production. DNA probes have also been developed for an intracellular parasites and bacteria infecting shrimp. DNA-based techniques will have an important role to play in efforts to develop sustainable shrimp culture in Asia and elsewhere. Production facilities in Thailand are currently using PCR techniques to screen shrimp post-larvae for WSSV. Culturing such larvae in closed (biosecure) or semi-closed culture systems can prevent or minimize viral infections, leading to a viable shrimp industry. The development of specific pathogen-free shrimp stocks will also depend on the use of such techniques.

The further development and use of DNA-based diagnostic techniques will also assist international efforts to control the introduction of exotic diseases into new geographic areas. Reliable and rapid techniques are needed by national and regional diagnostic laboratories to screen imported fish and shellfish for important pathogens. The Office International des Epizootics (OIE) or World Animal Health Organization, is a veterinary organization with 147 member countries. The OIE (through its Fish Diseases Commission) is responsible for tracking diseases of fish and shellfish that have a serious economic impact on aquaculture and capture fisheries. There is considerable potential to apply DNA-based methods for OIE testing if they can meet the stringent criteria of a standardized, validated, accurate, reliable and accessible diagnostic technique.

Impediments to the Use of DNA-Based Diagnostic Techniques

Although offering considerable potential, the routine use of DNA-based diagnostic techniques is hampered by a number of potential problems.

— The extreme sensitivity of these methods allows the detection of target DNA present at very low levels. However, positive results provide little quantitative assessment of the infection level, and do not indicate whether the pathogen is replicating or causing disease in the species tested. Thus, carrier status and viability of the pathogen are not determined using DNA-probes.

— The extremely high specificity of these tests, coupled with the ability of many viruses to rapidly change in genetic structure, can result in failure to detect a virus that has altered its genetic profile.

— Large differences in sensitivity are related to the PCR method used (e.g., 1-step PCR or 2-step PCR with nested primers).

— PCR methodologies are highly susceptible to contamination. Contamination during processing may result in false positives, particularly in 2-step PCR methods. PCR tests must be conducted in very well managed, clean laboratories.

— "False negatives" are easily caused by the selection of inappropriate host tissue sources for detection of the pathogen in question, incorrect choice of DNA extraction method, or low pathogen prevalence in the population sampled.

DNA-based detection and diagnostic methods have the potential for widespread application of in aquaculture. As the technology is already being adopted rapidly in developing countries in Asia, there is an urgent need to address these issues and to develop an action plan for research and training activities that will facilitate more effective utilization.

Shrimp Pathogens

Shrimp pathogens in the Asia-Pacific region presently listed by NACA include WSSV and YHV (notifiable), INNHV, GAV, MBV, BMNV and SMV (significant pathogens). HPV also may have significant effects on production. The range of histopathological and molecular techniques available for detection of these agents has recently been reviewed. Research needs pertaining to the application of DNA-based technologies vary, reflecting the stage of development of the technology and the relative importance of the pathogens. In view of their ongoing impact on shrimp aquaculture in Asia and OIE-notifiable status, this report has focussed primarily on the research status and needs for WSSV, YHV and viruses in the YHV complex. The urgency of implementing measures to control these viruses is presently the dominant concern for shrimp health in Asia.

WSSV. A number of PCR, nested-PCR and hybridization tests has been developed for WSSV detection. The tests use a range of different PCR primers and hybridization probes targeted to different and poorly defined sites in the WSSV genome. Methods of DNA sample preparation and PCR test protocols vary and there has been no objective comparison of the sensitivity and specificity of the tests. A recent analysis of PCR products amplified from WSSV samples obtained from a wide geographic range has indicated remarkable uniformity. However, the data suggested that at least one WSSV-related virus in the USA may be distinct. Other reports also suggest some variability in the detection specificity of PCR tests. There is evidence that a range of crustaceans and other arthropods test positive by PC, but the significance of these results for the epidemiology WSSV infection in shrimp is unclear. A more extensive analysis of WSSV variation should be conducted to determine the implications for detection and disease diagnosis.

The proliferation of PCR tests and protocols for WSSV detection also presents problems for comparative validation. The one-step and nested PCR primers and procedures described by Lo *et al.* are well documented in the published literature and the utility of the tests has been demonstrated for a range of applications. The test appears to be reliable and not subject to commercial restrictions, and would be suitable as a primary reference for standardization purposes. Estimation of the level of WSSV infection may have important applications in disease management strategies and this should be more clearly defined. It may be useful to adopt a quantitative PCR test as a secondary reference standard.

YHV. The YHV complex constitutes a group of related agents which includes yellow head virus (YHV), gill-associated virus (GAV) and lymphoid organ virus (LOV). YHV and GAV are closely related but distinct pathogens; LOV is a variant of GAV which occurs in healthy shrimp. Yellow head disease has been reported from many countries in the Asian region but, in most cases, the agents have not been clearly defined. To date, YHV has been shown to occur only in Thailand and GAV/LOV only in Australia.

An RT-PCR test is available for YHV but the test does not detect GAV. RT-PCR and nested RT-PCR tests are available for GAV. The GAV RT-PCR also detects at least some isolates of YHV but the test will not distinguish GAV and LOV. *In situ* hybridization probes have also been developed for YHV and GAV. The YHV probe detects both YHV and GAV. All tests developed to date have targeted sequences in the polymerase gene.

More research is required to determine the range and distribution of YHV complex viruses in the region, to identify other possible members of the complex and to develop both pan-specific and type-specific detection tests. If possible, tests should also be developed to differentiate pathogenic from non-pathogenic variants. There may be value in the adoption of multiplex PCR tests for YHV that allow discrimination of viruses in the complex.

Standardization of DNA-based Diagnostic Tests

A program of laboratory accreditation is proposed in order to achieve reliability and comparability of DNA-based test results in the Asia-Pacific region. The program should seek primarily to provide an improved regional capacity for effective disease control. However, if adequate and common levels of test performance can be obtained, the program may also provide a mechanism for eventual adoption of molecular-based methods for health certification by OIE.

It is proposed that FAO/NACA accreditation should be issued by reference laboratories designated for each pathogen. The reference laboratory should publish a detailed standard protocol including procedures for sample collection, DNA/RNA preparation, PCR reagent preparation and storage, PCR amplification and sample analysis, and preparation and use of control reagents. Adoption of the standard protocol will be facilitated by publication in local languages. The reference laboratories should also be responsible for:

— maintaining an agreed test as the primary reference standard against which other testing protocols should be assessed;

— maintaining standard PCR reagents including primers and suitable positive and negative controls;

— monitoring standards and providing technical advise to accredited laboratories;

— providing definitive diagnosis in difficult or unusual cases;

— retaining and archiving virus isolates for reference;

— ongoing assessment and Research and Development of DNA-based testing protocols.

Accreditation should initially focus on WSSV diagnosis. Because of the proliferation of WSSV PCR testing protocols, a primary reference standard should be identified. It is proposed that the one-step and nested WSSV PCR tests described by Lo *et al.* should be adopted as the primary standard. Implementation of a regional accreditation program should proceed following the issue of standard protocols. A relative evaluation of the diagnostic capabilities of participating laboratories should be conducted using standard coded samples of extracted DNA and shrimp tissue. An epidemiologist should be involved in the design of the evaluation and data analysis. The evaluation should be used as a basis for the development of an accreditation protocol. Standardization and accreditation for YHV diagnosis is considered to be premature, as there is presently insufficient information available on the relationship of viruses that constitute the YHV complex. The establishment of a WSSV accreditation program will facilitate future accreditation of laboratories for diagnosis of YHV and other shrimp pathogens.

Networking and Communication

NACA should be the conduit for maintaining communication between regional diagnostic laboratories, preferably through an email network. The establishment of an accreditation program should involve an initial meeting of participating laboratories to set up comparison protocols.

Training and Extension

There is a need for training both at farm-level and of diagnostic practitioners. Practitioner training should be through intensive short courses which include both theory and practice of DNA-based technologies and high quality tertiary courses for veterinarians and fish pathologists. Farm-level training is best achieved through in-country training of extension officers to meet local needs and should include: i) instruction on sample collection and preparation methods; ii) accurate interpretation of PCR results; iii) limitations of PCR technology; and iv) basic epidemiology.Standard procedures for the application of DNA-based technologies for disease prevention should also be documented. These should include the use of PCR in hatcheries and on-farm, and for the selection of SPF broodstock.

Mollusc Pathogens

With the possible exception of Australia and New Zealand, there is a lack of published or readily available information on molluscan parasites, pests and diseases in the Asia-Pacific region. Most countries lack dedicated expertise, facilities and infrastructures for molluscan health examination. However, in many countries, mollusc production (subsistence and aquaculture) is established and growing. Species under cultivation for food or secondary products include pearl oysters, edible oysters, mussels, scallops, abalone and capiz shell. Although there is no immediate application for use of DNA-based diagnosis methods on a routine basis for molluscs, tools already developed could be used for cross-checking the specific/generic identities of emerging parasites which appear related to known pathogens e.g. *Haplosporidium* spp., *Marteilia* spp. and *Perkinsus* spp. Unpublished data from Australia indicates the presence of *Haplosporidium* sp. in *Pinctada maxima*, and *Perkinsus* sp. in *Saccostrea commercialis*. In view of the presence of potential pathogens, the reported growth of mollusc aquaculture industries and the increasing pressure for live introductions and transfers, there is an urgent need for a survey of normal and diseased animals to obtain more extensive data on mollusc health.

There is also a need to establish national and regional expertise through training. As a prerequisite to implementation, NACA should contact National Coordinators to: i) confirm support for a mollusc health program; ii) identify commercially significant species; and iii) designate at least two technicians who would be dedicated to the project. NACA should also establish a panel of experts (from within and outside the region) who will support the project and provide reagents and reference material. Initially, training will aim to establish an initial baseline of expertise in husbandry, sampling, gross observations, fixation, anatomy, and histology. This first step would provide the basis for national programs for mollusc health monitoring. Where resources are limited, this could include market samples. Initial training would be followed (3-4 months later) by an advanced session for the same participants covering histopathology and technical problem solving. During this session, each participant would bring examples of their own material for joint consultation and evaluation. In addition, contacts with mollusc pathology specialists and reference material from other collections (slides and guides) should be available for these participants. The goal of this training program would be to establish a capability for independent monitoring of mollusc health in each country.

As a result of this training program, a clearer picture of the health status of molluscs and the diagnostic needs in the region should emerge. The connection between this initiative and the supporting network (comprising specialist mollusc pathology laboratories) will enable access to the advanced diagnostic tools discussed in detail in the full workshop forum (for finfish, shrimp and molluscan pathogens). The mollusc pathogen Focus Group determined that, in the Asia-Pacific region, these methods should be reserved for cross-checking material containing organisms resembling known pathogens for which advanced diagnostic methods are available in supporting laboratories (e.g.,*Marteilia* spp. DNA-probes). They could also be used to determine the geographic and host distributions of these pathogens as they emerge. DNA sequence analysis should also be conducted (e.g. to confirm positive probe results) to validate application of available probes to pathogens in the region. The need to develop additional molecular diagnostic tools and other pathogen/pathogen-group research needs will be determined as pathogens emerge from this surveillance program.

Finfish Pathogens

Diseases requiring application of DNA-based technologies.

There are 3 areas for which the introduction and use of DNA-based techniques is likely to have a significant impact on improving disease diagnosis in developing countries.

Mycobacteriosis. As there is a zoonosis risk associated with *M. marinum*, there is a special need to identify mycobacterium infections to the species level. Currently available PCR methods should be standardized and validated to the extent that stocks can be certified.

Viral nervous necrosis (VNN). VNN is a potential threat to an important grouper industry in the region. DNA probes are available to particular strains in Europe and possibly Australia. A comparative study of available probes is required and a validated method for screening wild grouper broodstock should be developed.

Epizootic ulcerative syndrome (EUS). EUS infections cannot presently be distinguished from occurrences of ulcerative mycosis in USA. There is a need for a standardized, validated *in situ* hybridization test that is specific for*A. invadans*. The development of PCR primers and/or hybridization probes for the EUS-associated rhabdovirus would also assist in understanding the disease syndrome.

Evaluation of Needs for Rapid Detection Techniques

The majority of pathogens causing the principal diseases of fish in the Asia-Pacific region can be detected and identified using existing methodologies. These include visual observation and light microscopy for parasitic infections, and routine bacteriology and histology for bacterial pathogens. Specific staining methods and antibody-based techniques are also utilized. Cell culture is used for grown and identification of viral pathogens. The primary fungal infection (EUS) is detected by histology. However, DNA-based technologies would assist in controlling the spread of fish diseases disease if used for confirmation of diagnosis and for screening fish for specific pathogens such as *M. marinum*, VNNV and *A. invadans*.

Status of Research

Research is already underway on all three areas:

Mycobacterium spp. Monoclonal antibodies (MAbs), PCR primers and DNA probes are available. The MAbs can discriminate *M. marinum* at species level but do not detect all isolates. New MAbs are currently being developed. Two PCR-based methods have been developed, one employing enzyme restriction analysis of the PCR product, and the other employing reverse cross-blot analysis. The later appears to be more specific and can detect all 3 species infecting fish (*M. marinum, M. fortuitum* and *M. chelonei*). This test is presently used to detect aquatic mycobacteria in fish tissue samples, water samples and in human clinical biopsies but it requires further standardization and field validation.

VNNV. VNN is present in Australia and has recently been detected in grouper in Thailand. Diagnostic antibodies, PCR primers and DNA probes have been developed for VNNV isolates from the Mediterranean region where the disease causes widespread mortalities. These reagents should be used to determine their specificity for VNNV isolates from the Asia-Pacific region.

EUS. PCR and in situ hybridization tests are currently being developed to detect *A. invadans* from Asia. These should be utilized to screen samples from different geographical locations for the presence of *A. invadans*.

Primers should also be developed against the rhabdovirus which is often isolated in association with EUS.

Key Constraints to Establishment of International Standards

The need for a more formal set of protocols and procedures for PCR-based diagnostic assays has been recognized for a number of years. However, as in many other areas of diagnostics, there has been no serious attempt to set agreed international standards. For a technique with the potential power PCR, the lack of clear guidelines for the correct performance is a serious constraint to its routine use. A second and more serious constraint is the almost complete absence of validation studies undertaken on field samples for any of the currently available assays. Although validation programs are expensive and time consuming, and carefully designed laboratory studies can provide much of the data required to ensure that PCR-based assays perform with acceptable precision, information on the meaning of the results generated by non-culture-based diagnostic techniques can only be obtained through field validation programs. Therefore, collaborative projects to assess currently available PCR-based diagnostic techniques through comparative and predictive field validation studies are urgently required and should be actively supported by funding agencies.

Recommendations for Research Programs

Adoption of PCR-based diagnostic assays, to replace more established methods is an attractive option for many laboratories. These assays are seen as more sensitive, rapid and logistically simple, and have the allure of high technology. However, in the absence of adequate internal (laboratory) and external (field) validation, interpretation of the meaning of the results generated by the PCR-based assays will remain problematic. While an assay may be acceptable in a research context it will not be acceptable in either a diagnostic or regulatory context and could generate misleading data. Therefore, the priority of any research program that aims to replace current diagnostic methods with PCR-based assays must be a carefully considered validation program that addresses the issue of interpretation of meaning.

As a first approach, comparative validation programs should be conducted in which PCR-based assays are performed in parallel with established methods over a reasonable period of time and using statistically significant number of samples. This type of validation program would also allow standardization of the performance of the assay in the laboratory. Standard reference material (both positive and negative controls) supplied from a central reference laboratory would be important to ensure the precision of the results generated by such a study.

More importantly, it is essential that results generated by PCR-based diagnostic assays can be related to field situations. Positive or negative results generated by the assay should be reliably related to actual disease episodes. It is only through such 'predictive validation' that interpretation of the results in relation to disease diagnosis is meaningful. Some components of a predictive validation study could be conducted retrospectively if field data of adequate quality were available.

International Standards for Disease Diagnosis and Health Certification

In 1995, the World Trade Organisation (WTO) implemented the Agreement on the Application of Sanitary and Phyto-Sanitary Measures (SPS Agreement) to define the conditions under which countries can impose sanitary conditions on imports of animal and animal and plant products. This Agreement is intended to promote international trade by requiring all member states to use only international standards for reducing disease risks associated with imported products. For animals and animal products, the Agreement identifies the guidelines and recommendations established by the Office International des Epizooties (OIE) as the appropriate and preferred international standard. However, the Agreement also allows governments to use standards developed by other relevant international organisations whose membership is open to all WTO members, or to use higher standards if an appropriate risk assessment provides adequate scientific justification.

In Europe, EU Member States have agreed common standards for the health conditions applying to intra-Community trade in aquaculture animals and their products, based on those recommended by OIE. Countries in the Asia-Pacific Region may also wish to agree on common standards of diagnostic and health assurance tests. However, if international standards established in a particular region differ significantly from those recommended by the OIE, trade with countries outside the region may be adversely affected. For example, if the standard adopted in a region is *lower* than the OIE standard, countries which apply the OIE standards could rightly refuse imports on the basis that they have insufficient health guarantees.

Conversely, if the standard agreed in a region is *higher* than those of OIE, the same standard must be applied to imports from other regions, to ensure equivalence and protection of the regional health situation. Furthermore, if an individual country adopts higher health protection standards, the SPS Agreement requires scientific justification based on import risk assessment. For diseases not listed by the OIE, trading countries may agree on any mutually acceptable standards.

If a regional training and laboratory accreditation program for molecular diagnosis is to be established in the Asia-Pacific region, the adoption of tests recommended in the OIE Diagnostic Manual for Aquatic Animal Diseases would assist in achieving universally agreed standards. However, as the Manual is revised only every 3 years or so, better methods may emerge from research before the next edition is published. In considering a case for adoption of new molecular diagnostic methods, the suitability must be rigorously assessed in consultation with independent experts. Acceptance is more likely if the method has been published, has received wide scientific acceptance and has been standardised, and preferably validated, in comparison with other standardised methods.

Key Outcomes and Recommendations

There is considerable scope for more effective use of DNA-based methods of pathogen detection and disease diagnosis in Asia-Pacific aquaculture. However, implementation of standardised practices that produce reliable, useful and comparable data will require a significant investment in research, training and infrastructure development. Effective implementation will also be assisted by enhanced communication between aquatic animal health practitioners in the region and scientists with expertise in disease diagnosis and pathogen detection.

Although there are some common themes, it is also evident that there are significant differences in the current relevance of DNA-based methods of pathogen detection for the different aquaculture sectors. DNA-based methods are particularly suitable for detection and diagnosis of shrimp and mollusc pathogens because of the absence of an antibody response in invertebrates and lack of suitable cell lines for virus cultivation. In shrimp, the primary pathogens are well known and many DNA-based methods have already been developed. However, in molluscs there is very limited knowledge of pathogens and few diagnostic procedures of any kind are being employed in the Asia-Pacific region. In fish, antibody and culture-based diagnostic methods are available and considered to be robust and effective for routine diagnostic applications. As such, DNA-based methods in fish appear to be most suitable for confirmatory diagnosis and rapid screening of low level or unapparent infections. To achieve maximum impact, it is essential that research and training programs recognise these differences and are tailored to reflect current levels of knowledge and sector-specific needs.

Where DNA-based tests are available and/or suitable, the most significant impediment to effective implementation is the lack of standardised methodologies that are validated for specific applications. There is a need for international agreement on methodologies that have been rigorously evaluated and accredited for specific applications in disease diagnosis and pathogen screening. There is also a need to ensure that tests are performed by trained staff with access to standardised reagents and suitably equipped laboratories.

Because of existing limitations on the reliability and accessibility of the methods, international standards recommended by OIE do not presently include DNA-based methodologies. However, the potentially high sensitivity and specificity and relatively low cost of these tests has resulted in a surprisingly rapid adoption rate in Asia, particularly for shrimp pathogens. Therefore, it is essential that DNA-based tests are assessed on their merits against existing technologies and that programs to achieve improved performance and international standardisation should be developed.

References

Ambrosia, R.E. and De Wall, D.T. 1990. Diagnosis of parasitic disease. *Reviews of Sci Techn., Office Intern. Epizool.* 9, 759-778.

Knowles, D.P. and Gorham, J. R. 1990. Diagnosis of viral and bacterial diseases. *Reviews of Sci. Techn., Off. Intern. Epizool.* 9, 733-757.

Lightner, D.V. and Redman, R.M. 1998. Shrimp diseases and current diagnostic methods. *Aquaculture* 164, 201-220.

Plumb, J.A. 1997. Trends in freshwater fish disease research. p. 35-47. *In:* T.W. Flegel and I.H. MacRae (eds.) *In:Diseases in Asian Aquaculture* III. Fish Health Section, Asian Fish. Soc., Manila.

Rana, K.J. 1997. Recent trends in global aquaculture production: 1984-1995. 14-19p. FAO *Aquaculture Newsletter.* No. 16. August 1997, FAO, Rome. 28pp.

10

Current Status of Aquaculture Biotechnology

Capture fsheries and aquaculture supplied the world with over 113 million tonnes of food fsh in 2007, providing an apparent per capita supply of 17.1 kg (live weight equivalent), which is among the highest on record. Global production of fsh from aquaculture has grown rapidly during the past four decades, contributing signifcant quantities to the world's supply of fsh for human consumption. Aquaculture currently accounts for nearly half (44.3 percent) of the world's food fsh (Figure 1). With its continued growth, it is expected that aquaculture will in the near future produce more fsh for direct human consumption than capture fsheries.

Started as primarily an Asian freshwater food production system, aquaculture has now spread to all continents, encompassing all aquatic environments and utilizing a range of aquatic species. From an activity that was principally small-scale, non-commercial and family-based, aquaculture now includes large-scale commercial or industrial production of high value species that are traded at national, regional and international levels. Although production remains predominantly Asian and is still largely based on small-scale operations, there is a wide consensus that aquaculture has the potential to meet the growing global demand for nutritious food fsh and to contribute to the growth of national economies, while supporting sustainable livelihoods in many communities.

In 2006, fsh provided more than 2.9 billion people with at least 15 percent of their average per capita animal protein intake. The contribution of fsh to the total world animal protein supplies grew from 14.9 percent in 1992 to a peak of 16.0 percent in 1996 before declining to about 15.3 percent in 2005. Notwithstanding the relatively low fsh consumption in low income food defcit countries of 13.8 kg per capita in 2005, the contribution of fsh to total animal protein intake was signifcant – at 18.5 percent – and is probably higher than indicated by offcial statistics in view of the under-recorded contribution of small-scale and subsistence fsheries and aquaculture.

Aquaculture is the farming of aquatic organisms including fsh, molluscs, crustaceans and aquatic plants. Farming implies modifcations and intervention in the production cycle such as regular stocking, sorting, feeding and protection from predators in order to enhance production.

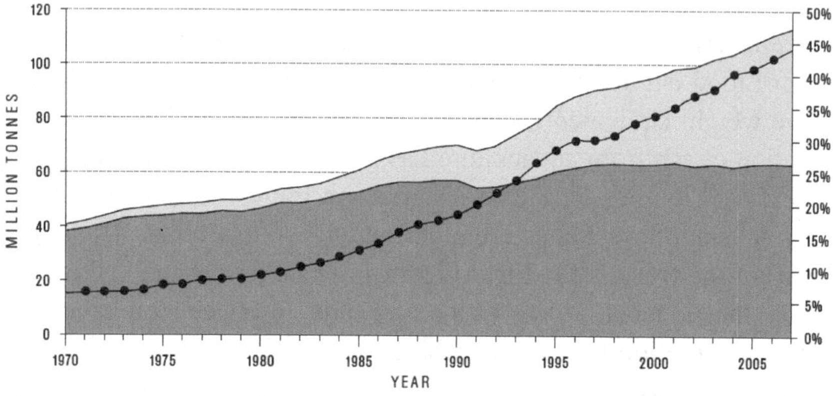

Figure 1. Contribution of food fish supply from capture fisheries and aquaculture

It is important to note that aquaculture has a long tradition in the developing countries of the Asia-Pacifc region, supplying most of the world's aquaculture production (over 90 percent), and making important contributions to the livelihoods and subsistence of small-scale farmers and coastal populations in many countries in the region. In Latin America, small-scale aquaculture has yet to be widely developed; however, there are several examples of newly established industries based on intensive aquaculture practices, especially using exotic species. Salmon farming in Chile is one of the best examples, but there are also expanding aquaculture industries for shrimp and tilapia culture in Ecuador, Costa Rica and Honduras. While Europe and North America import signifcant quantities of farmed aquatic animals, they also produce fsh and shellfsh both from freshwater and marine environments. Africa's contribution to global aquaculture is still small; however, the region is moving forward and increasing production.

Aquaculture covers a wide range of species and methods. It is practised from the cold waters of the far north and south, where fsh like salmon, Arctic char and sturgeon are grown in ponds, fowing raceways and cages in the sea, and through the latitudes as far as the tropics, where carp and tilapia fourish in freshwater and shrimp and sea bass are farmed along the coasts. It ranges from the production of fsh in naturally occurring ponds in rural areas to the intensive culture of ornamental fsh in plastic tanks in the middle of a city. It is practised by the poorest farmers in developing countries as a livelihood and supply of much needed protein for their families, and by urban sports shop owners in Europe and North America producing baitfsh for weekend anglers.

Aquaculture systems can range from an intensive indoor system monitored with high tech equipment through to the simple release of fry and fngerlings to the sea, but the aim remains the same: to improve production. Some of the simplest production systems are the small family ponds in tropical countries where carp are reared for domestic consumption. At the other end of the scale are high technology systems such as the intensive indoor closed units used in North America for the rearing of striped bass or the sea cages used in Chile and Europe for growing salmon and bream.

All products and systems are geared to produce animals for market and are much governed by market demand at all levels. Regardless of whether it is a high value commodity like shrimp, salmon or grouper, or a low-value commodity such as carp and Tra catfsh, all products are destined for markets, be they local, regional or international. All production systems contribute to food security and human development although small-scale rural production systems provide more support to improving or maintaining livelihoods and generating employment and income for many around the world.

It is important to note that most of these small-scale aquaculture activities occur in developing countries, especially in regions or rural areas where food supply is at risk. For example, tilapia has become a globally important aquatic species that is produced in nearly 100 developing countries worldwide. According to FAO, about 80 percent of the world's farmed tilapia comes from small-holders in developing countries, and this species is particularly prominent in production systems in the Asia-Pacifc, the region that provides most of the world's aquaculture supply.

Another good example of extensive aquaculture is the production of major carps in India. In this case, the majority of the production takes place in rural areas with relatively few impacts on the environment, particularly by using multitrophic culture of species such as catla (*Catla catla*), rohu (*Labeo rohita*) and mrigal (*Cirrhinus mrigala*). It is true that some instances of uncontrolled aquaculture development have caused signifcant negative environmental and social impacts. However, except for a very few species, there are few negative environmental impacts associated with current production systems and practices. Moreover, most traditional and extensive systems produce fsh with little or no negative environmental or social impact.

There has been a steady increase in the growth of aquaculture in developing countries, the rate of growth being twice that of developed nations. The most recent fgures for global aquaculture production show that more than 90 percent of total fsh production comes from developing countries, particularly China which contributes about 70 percent of the total global fsh and shellfsh production. Aquaculture is thus often one of the most important food production sectors in developing countries, and in many cases it is one of the most important sources of both food and income for rural populations.

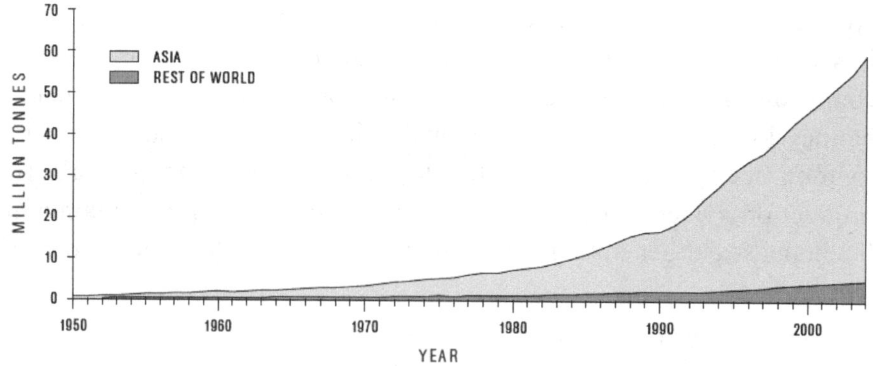

Figure 2. Global aquaculture production

Aquaculture practice is an example of a strong continuum of production systems. From the simplest production system with absolutely no inputs and with minimal interventions, aquaculture ranges up to highly sophisticated, fully automated, industrial production systems comprising submerged offshore cages producing large quantities of fsh from a single unit. Intensive or extensive aquaculture requires good quality seed for farming. Seed quality is not only dependent on good hatchery technology, but also on good broodstock with improved genetic quality. The genetic quality of the broodstock and seed used in aquaculture can be improved using biotechnological tools and procedures. There have been some interventions, and good results have been reported.

Modern aquaculture, through the intensifcation of culture systems and the diversifcation of both the species cultured and the culture methods employed, often creates an ideal environment for disease-causing organisms (pathogens) to fourish. The expanded and occasionally irresponsible global movement of live aquatic animals has been the cause of transboundary spread of many pathogens, which have sometimes resulted in serious damage to aquatic food productivity. Some of these pathogens have become endemic in culture systems and in the natural aquatic environment, thus making them diffcult to eradicate. Since they have become endemic, recurrent pathogen incursions and disease outbreaks occur in farms making it diffcult for the farmers to effectively manage farm health. Instead of implementing effective health management strategies and practices, many farmers opt to use antimicrobials as treatments. There is therefore a need to develop alternate methodologies and tools for maintaining aquatic animal health in aquaculture systems. Such tools and methodologies are generally the result of biotechnological research and several success stories exist. Similarly, biotechnological research has also helped in the improvement of feeds, feeding and nutrition as well as of water quality and the environmental impacts of aquaculture.

Main Areas where Biotechnologies being Applied

Genetic Improvement and Control of Reproduction

Aquaculture is still the fastest growing food producing sector, compared with other food commodities. One of the reasons for this is the diversity of species in culture at present, and the genetic diversity that can be exploited through captive breeding and domestication, enabling the development of improved culture methods for a diverse array of species to expand commercial aquaculture. A lack of knowledge of the biology of many of these species and the cost of technology development are constraints that explain in part why biotechnologies are only now emerging as useful tools for increasing the productivity and sustainability of this sector. Aquaculture is a sector that is likely to beneft greatly from the application of appropriate genetic and reproduction biotechnologies to increase food production.

Despite the current trend towards the intensifcation of production systems, aquaculture has not made full use of conventional technologies such as genetic selection and breeding improvement programmes to increase production as have other food production sectors. The rearing of many newly cultured species is to a large extent based on juveniles and/or broodstock obtained from the wild. In order to establish practical breeding programmes to produce seed in hatcheries it is necessary to have a detailed understanding of the complete production cycle. Such knowledge is also required in order to disseminate breeding improvements to the production sector.

One of the best examples is the inability to fully domesticate *Penaeus monodon*, the black tiger prawn which is arguably the most valuable species produced globally. Although specifc pathogen-free (SPF) hatchery stocks bred for improved growth have become available recently, production still depends on broodstock collected from the wild. As a result, production of this species has been replaced over the last few years by that from the white shrimp, *L. vannamei*. Improved SPF *L. vannamei* have been readily available for some time and now supply essentially all farmed white shrimp and more than 60 percent of all farmed penaeid shrimp world wide. The shrimp aquaculture sector therefore illustrates the benefts of genetic improvement for increasing production and the competitiveness of aquaculture industries.

The *P. monodon* example illustrates how a lack of knowledge concerning some phases of the life cycle such as reproduction or metamorphosis may be a limiting factor in developing domesticated stocks. Certain species of tuna, a marine resource that is being harvested under a quota system, are now produced in considerable quantities in captivity or culture. The aquaculture production of this valuable species will undoubtedly increase once the life cycle is closed and the hatchery production of tuna fry becomes a reality. This scenario is also applicable to the hatchery production of mollusc species. There is a huge demand for spats (fertilized shellfsh larvae) but most spats are still coming from the wild.

The use of hormones for the control of reproduction has been primarily developed for inducing the fnal phase of ova production, i.e. for synchronizing ovulation and for enabling broodstock to produce fsh in the frst part of the season or when environmental conditions suppress the spawning timing of females. These procedures began with the pioneering work of Houssay, who demonstrated that extracts of the hypophysis (pituitary gland) can have an effect on sexual maturation of fsh and reptiles. These results allowed the development of a relatively simple procedure consisting of injecting hypophyseal extracts purifed by chromatography that contain products such as inductive hormones related to sexual maturation. Human chorionic gonadotrophin and the gonadotrophin-releasing hormone (GnRH) were also used to control the maturation of many fsh species without limitations due to species-specifc effects. GnRHa, an analogous GnRH developed chemically, is more effcient in inducing maturation and is relatively inexpensive. It can be injected or administered by means of pellet implants which facilitate its practical use.

The use of hormones such as GnRHa has allowed advancement of the date of egg-laying in several species of fsh, mainly salmonids, although for relatively short periods of time. Several other molecules are currently under development for use in molluscs (e.g. scallops, oysters and mussels), where synchronous reproduction is required for the hatchery rearing of larvae for aquaculture production in developing countries instead of using seed obtained from natural banks.

Biosecurity and Disease Control

Disease outbreaks are a serious constraint to the development of intensive aquaculture systems and can have a major impact on production due to mortality and decreased growth. It has been recognized that disease is the most signifcant factor impacting the intensive production of shrimp, salmon, carp and tilapia, with losses of 10-90 percent of total production. Although many aquatic animal pathogens are well studied, unlike in terrestrial animals the spread of pathogens is easy through water and control is diffcult due to high density culture in fuid environment. Disease occurs in all systems, from extensive to intensive, although heavy losses are always possible in intensive production systems.

Intensive and semi-intensive aquaculture can have important effects on the quality of the aquatic environment in which the animals are reared. Poor water quality resulting from increased waste products, inadequate farm management, increased stocking densities within farms and increased densities of aquaculture units per sector can increase the likelihood of disease outbreaks and other environmental problems such as eutrophication, episodic oxygen shortages, algal blooms etc., all of them potentially resulting in high mortalities. A more "systems-oriented approach" is therefore needed to provide suitable husbandry for effective growth and to control disease outbreaks effectively.

There is a greater need for management intervention in intensive systems. Here biotechnological tools can be a valuable part of management approaches. Their scope of application is broad – they can be used as sensors in the production environment, for waste management (through controlled microbial technologies), and for disease detection and control (molecular methods). Traditionally, disease control is often carried out only after mortality has been observed. In the past, the diagnosis of fsh diseases has been achieved primarily using histopathological methods supported by parasitological, bacteriological and viral studies based on necropsy and *in vitro* cell culture. These are well-proven techniques. However, they require a high level of expertise and are often quite time-consuming, not being amenable to automation. For these reasons, although expert training is required, polymerase chain reaction (PCR) technology (described later) has become an important tool for pathogen assessment in developing countries, for example in the shrimp industries of Asia and Latin America.

Environmental Management and Bioremediation

Aquaculture has often been accused of being unsustainable and not environmentally friendly. Although in some cases, where aquaculture development has failed to live up to the global expectations of sustainable development, these allegations are not entirely unfounded, the majority of aquaculture is practised sustainably and with a high degree of environmental conscientiousness. Reducing the impact of effuent discharge, improving water quality and the responsible use of water are key areas to be considered during aquaculture development. A number of biotechnologies are being used to address these areas: bioremediation for the degradation of hazardous wastes; the use of vaccination and probiotics to reduce antimicrobial use;

and the use of DNA-based methodologies for the early detection of toxin-producing algae.

Biodiversity Conservation and Fsheries Management

In fsheries management, conservation is an important concept. Good fsheries management requires effective conservation measures, which require better understanding of the population structure of the fshery. One of the most important population parameters for assessing the fate of a population is the effective population size (N_e), which determines the amount of genetic variation, genetic drift and linkage disequilibrium in populations and can be calculated as half the reciprocal of the rate of inbreeding. There is much concern in fsheries and aquaculture production about the potential loss of genetic variation that may result from the relatively high rates of inbreeding expected in these populations. This is because many fsh and shellfsh species produce thousands or even millions of fertile eggs from a single female. Due to differences in the biological and environmental factors affecting the survival of individual families, many species show a relatively large variance in family size, further decreasing the N_e. Fisheries resource managers have focused on the actual number of individuals in a population (census numbers), which may be many times higher than the $N_{e.}$ Therefore, it is diffcult or even impossible in some cases to infer the N_e using the census number. Inadequate procedures for stock enhancement can yield a very small effective population size due to the high prolifcacy of fsh and shellfsh species. Thus, a very small number of breeders could be used for restocking purposes, and bottlenecks can affect the ftness of the population in future generations. A range of biotechnology-based approaches are being used to conserve wild fsh populations such as the use of molecular markers: to estimate N_e in wild populations; to study gene fow between farmed and wild fsh populations; and to monitor and understand changes in wild fsh population sizes.

Applications of Biotechnologies in Developing Countries

In fsheries and aquaculture, although perhaps not as much as in livestock and crop production, some biotechnologies have been used in developing countries. As mentioned earlier, use of biotechnologies in fsheries is very limited whilst in aquaculture biotechnologies are represented in a few felds such as genetic improvement, disease control, feeds and nutrition and environmental improvement.

Genetic Improvement and Control of Reproduction

Polyploidy

Many fsh and shellfsh species are relatively tolerant to chromosomal manipulation in the early stages of their development. The use of genetic manipulation including polyploidy (i.e. increasing the number of sets of chromosomes) to improve aquaculture production has been examined. However, there has been little discussion of the use of these technologies in practical management programmes in developing countries or on how they can be used effciently within the context of breeding programmes. Furthermore, the potential value of this technology under practical conditions for enhancing the performance of commercial populations in developing countries is not clear.

The induction of polyploidy has been considered by many researchers because of the advantages related to triploid sterility. For example, triploids (with three sets of chromosomes) may be useful for conservation programmes where sterility can prevent introgression of genes from escaped individuals of commercial stocks into natural populations, or in commercial operations where sterile fsh are desirable to prevent side effects such as deterioration of carcass quality due to maturation or the occurrence of high mortalities in stocks when males mature early or that occur prior to maturation, especially in populations of Pacifc salmon.

Triploidy leads to the production of nearly completely sterile populations, as has been observed in rainbow trout populations with spontaneously occurring triploids. However, the degree of reproductive disruption varies depending on the species and the sex. Gametogenesis is severely disrupted in triploid females of salmon while, in contrast, triploid males usually display secondary sexual dimorphism (i.e. darkened skin colour and modified body conformation), courtship behaviour and develop an endocrine profle similar to that of diploid males. Spermatogenesis, however, appears to be somewhat reduced in comparison with diploid males. Although triploid males are to a great extent sterile, fertilization has been reported to occur. In the salmon aquaculture industry, sexual maturity and the associated gonadal development is generally an economic drawback as metabolic energy is diverted from somatic cell growth to reproduction, resulting in the deterioration of fesh quality and appearance. In this situation, the advantages of triploidy occur primarily after the onset of maturation when triploid female fsh may show an extension of growth and the inhibition of maturation prevents the normal degradation in carcass quality that is observed during the spawning season. Furthermore, female salmon triploids show a signifcantly higher dress-out percentage and higher pigment (canthaxanthin) retention, but concomitantly, there is an increase of fat deposition surrounding the viscera.

In developing countries, the practical implementation of triploidy in fsh production has not been very successful. Most of the research on the application of this biotechnology has been experimental, without extensive testing under practical conditions that consider the wide range of environments in which aquaculture takes place. In species such as tilapia and carp, testing of triploidy is a very important issue considering that there is intraspecies variation in the rate of triploidization due to the size and quality of the eggs. For this reason, it is not possible to ensure 100 percent triploidy when applying this technique on a commercial scale. Also, an increased mortality rate at the beginning of the life cycle and the detrimental effect of triploidy on growth and ftness could be signifcant constraints to the commercial production of triploids in some species. The lack of knowledge about the effects of competition between triploids and diploids in large extensive conditions in species such as tilapia could also be a disadvantage, since triploids sometimes lack robustness compared with normal diploids, but this expression varies among species. In many cases, the variation in performance between diploid and triploid stocks has not been fully estimated, and thus it may not be possible to accurately predict the relative performance of triploids in commercial conditions, which may be a problem in conventional breeding programmes of many fsh and shellfsh species.

In developing countries, for various reasons, these techniques are not currently used for commercial purposes. Tilapia, for example, cannot be easily reproduced using external fertilization which is a prerequisite for shock treatment. Furthermore, when a very small number of eggs are obtained per spawn, it is not possible to ensure a constant rate of triploidy per spawning. In rainbow trout, it is only proftable to use triploid females since males show some degree of reproductive onset. For developing such female triploid populations, neomales (i.e. morphologically male but genetically female) are required, which in some instances are diffcult to stock up to a commercial scale. In Indian carps, sterility aiming at faster growth and thus enhanced production may not be cost-effective since harvesting after one year of age is not proftable (males mature at one year of age and females when approaching two years).

In southern India, precocious maturation is a potential constraint on yields of cultured common carp as both males and females can attain sexual maturity well before reaching a marketable size. However, triploid fsh did not show any improvement over diploid individuals except for higher dress-out percentages.

Despite the plethora of research conducted on triploidization and chromosomal biotechnologies, there remains a gap between research fndings and the practical implementation of triploidy. Several reasons explain this fact. The usefulness of applying chromosomal biotechnologies such as triploidy for aquaculture production seems to be very species-specifc, and therefore in some cases (such as in salmon, tilapia and carp), the advantages due to delayed maturation or increased growth are unclear. Furthermore, the results of using these techniques to increase growth rate or delay reproduction are not seen as suffciently benefcial for the technique to be implemented on a large scale.

For the technology to be practical, it should be possible to produce all-triploid populations without the need to test the triploidy status of each batch of embryos produced. Because triploidy induction using thermal shock is not 100 percent effective, this is a serious drawback to the large-scale commercial application of the technique. Crossing between tetraploids (with four sets of chromosomes) and diploids is a way to produce 100 percent triploids; however, in most species tetraploid production is not straightforward. Furthermore, the genetic lag between the tetraploid population and the diploid breeding programme can seriously affect the effciency of the production system. For all these reasons, this technology has not been used extensively in developing countries for production purposes.

Gynogenesis/androgenesis

Gynogenesis is the production of an embryo from an egg after penetration by a spermatozoon that does not contribute genetic material. Androgenesis is the production of an embryo from an egg whose DNA was inactivated and which was fertilized using normal sperm. In both cases, the diploidy is restored using heat/cold shocks. In gynogenesis, if diploidy is restored soon after fertilization, the procedure is called meiotic gynogenesis due to the fact that the second polar body is retained, and this procedure is similar to what is expected under autofertilization in terms of inbreeding. If shocks are applied later or in androgenesis where the ova were DNA-irradiated for DNA inactivation, the same chromosome is duplicated and thus the embryo is a double haploid individual which is completely inbred for every locus.

Several papers have discussed the usefulness of this type of reproduction for genetic analysis in carp, tilapia and rainbow trout breeding programmes. In some cases, the use of gynogenetic individuals has been suggested for capitalizing on non-additive genetic effects to increase additive genetic variance and for product uniformity. However, the production of gynogenetic lines is not without problems. After a frst round of gynogenesis from an outbred population, deleterious and/or lethal effects can be fully expressed in the double haploid progeny, which may be a problem when implementing a breeding programme from this source. Furthermore, phenotypes cannot actually be a direct refection of the same trait measured on normal progeny due to developmental instability. Therefore, the utility of this type of reproduction for practical use in breeding programmes is seen as risky in most cases. Nonetheless, they can be used effectively for developing powerful quantitative trait locus (QTL) mapping experiments using the surviving clonal lines of this sort obtained from an outbred population, but this requires having available the gynogenetic lines that are needed for further assessment.

Controlling time of reproduction in fsh and shellfsh

So far, the application of hormonal treatment has been quite successful especially for controlling reproduction in broodstock. This is particularly the case in salmon and trout farming in Chile where either implants or injection of the hormonal compound are used extensively in salmon farming for synchronizing reproduction. Since hormone application is not done in the commercial fsh, but rather in the broodstock which are discarded for human consumption, these procedures are not subjected to a negative consumer preference. In carp breeding, the use of hormones has made it possible to artifcially manipulate the number of times and the timing of spawning of major Indian carps and African catfsh.

Development of monosex populations

One of the major constraints in practical programmes in developing countries is the fact that mixed sexed populations can behave poorly in production conditions. This is primarily due to the negative side-effects of early reproductive onset that decrease the growth rate through a series of physiological mechanisms. The faster growth rate of the other sex is probably caused by its later maturation. The negative relationship between growth rate and gonadal development has been found in many species. One explanation of this fnding is the appearance and accumulation of sex hormones that act as growth inhibitory agents.

The advantages of monosex culture depend on the species involved. This is because one sex may be superior in growth or have a more desirable meat quality, or to prevent reproduction during grow-out or the appearance of sexual/territorial behaviour (aggressiveness) that occurs when a mixed sexed group triggers the reproductive season. For example, female sturgeon are more valuable than males because they produce caviar; female salmon are more valuable because sexually precocious males die before they can be harvested, and salmon roe has an economic value; and male tilapia are more desirable than females because they grow twice as fast and because reproduction is not signifcant in males during grow-out.

The sex of fsh can easily be manipulated using hormonal treatments. In many fsh and shellfsh species, sex is not permanently defned genetically and can be altered by a number of factors including hormonal treatment during the early stages of development. Gonadal development starts from primordial germ cells, with females starting differentiation prior to males. The point in time when differentiation occurs depends on the species involved. In tilapia and trout, this mechanism is triggered early in life, while in grass carp and paddlefsh it is the opposite. Considering this pattern of development, treatment with the steroid methyl testosterone can be used to develop all-male tilapia populations and androgens (male sex hormones) can be used in trout and carp monosex culture.

There has been concern about the use of hormones in animal production including in aquaculture systems, arising from the risk of presence of residues in fnal products. Although there is little evidence regarding hormonal residues in fsh whose sex has been reversed early in life, consumer acceptance may be compromised as a result of the perception of hormonal treatment itself. For this reason, it appears that other biotechnologies have had more use in those developing countries whose production goes mainly to export markets.

A variation on this scheme is to produce all-male progeny in one more generation. This requires feeding young fsh with estrogens (female sex hormones), resulting in a population of all-female fsh. These morphologically female but genetically male fsh (neofemales) are then raised to maturity when they are mated to normal male fsh. After maturation, the all-male fry produced are tested in order to identify the "super males" (YY), which are then crossed to normal females (XX), thus generating all true male (XY) progeny. The importance of this method is that male fry for commercial production can be produced that have never been treated with hormones. However, one of the disadvantages is that this technique requires more than a single generation to obtain the all-male fry, i.e. this procedure cannot be used without extensive progeny testing to determine which "female" fsh will produce all-male progeny, thus requiring a reasonable time span for developing the neomales.

Although tilapia breeding programmes using YY super males are possible, this procedure is not necessarily required because the application of direct hormonal treatment of undifferentiated fry to produce monosex populations is still a major breakthrough.

However, the great expansion of tilapia aquaculture in Asia has been due to mixed-sex tilapia culture which addresses the high demand for relatively small fsh (i.e. fsh less than 300 g) that can be obtained by rearing the highly selected genetically improved farmed tilapia and other strains.

Cryopreservation

The aim of the cryopreservation of gametes is related to:

— disseminating semen from males obtained from selection programmes showing signifcant response;

— "refreshing" commercial populations in order to avoid the negative impact of bottlenecks;

— directly assessing the rates of genetic gain in ongoing breeding programmes;

— making semen available across the reproduction window when asynchrony of reproduction exists between males and females (usually males mature earlier than females).

Sperm cryopreservation has been successfully implemented for a number of cultured fnfsh and shellfsh species, and modest success has been achieved in the cryopreservation of shellfsh embryos and early larvae. Cryopreservation of fnfsh ova and embryos has not been successful, which is a major difference with respect to terrestrial animals. This is mainly due to the size of the ova which are usually large and have thick chorionic membranes that do not facilitate the inclusion of cryoprotectors.

The use of cryopreserved gametes for commercial purposes is still very limited in developing countries. One explanation is that this biotechnology may require specialized labour and automated procedures to decrease variability in success rates among batches of sperm. Furthermore, it is still uncertain whether this method is economically advantageous compared with disseminating improved broodstock using larval material. In spite of this, the technology has been used for disseminating improved "Jayanti" rohu in India and for the dissemination of improved semen in Sri Lanka (P. Routray, personal communication). In rainbow trout, cryopreservation has been used for storing semen from neomales, but the problem of highly variable fertilization success remains.

Genomics

Genomics is the study of the genomes of organisms. It includes the intensive efforts to determine the entire DNA sequence of organisms via fne-scale genetic mapping.

Genome sequencing

One of the major constraints in the rearing of many different aquaculture species is the lack of adequate genomic information. This is because sequencing all the species currently used in aquaculture would be costly. Productive species currently being sequenced are the tilapia and the Atlantic salmon (*Salmo salar*). A multinational initiative for Atlantic salmon aims to sequence the genome using the Sanger method to obtain a coverage of more than six-fold.

The project is a partnership between Canada, Norway and Chile, countries that are interested in applying this sequence data for studies related to enhancing conservation and production. The project's output will be delivered to the public domain and provide the required genomic resources for developing single nucleotide polymorphism (SNP) chips that will help implement marker-assisted selection (MAS) programmes in Chilean salmon aquaculture.

Functional genomics

The recent availability of massive amounts of information from functional genomics such as microarrays that are used to assess gene expression or sequence polymorphisms has contributed signifcantly to the genomic biotechnology in aquaculture. Two colour microarrays have been developed for salmonid species that are publicly available and are currently used to assess disease resistance traits in salmon and for candidate gene discovery. In shrimp, several platforms have been devised in China, Australia, Taiwan Province of China, Singapore and also the United States.

The main use of this resource has been to study differential expression of the transcriptome after viral or bacterial acute infection, but also as bioindicators for assessing chronic disease response. Microarrays are being applied to the felds of ecotoxicology and nutrigenomics. For example, gene expression analysis has been used for assessing the effect of pre-challenging white spot syndrome virus (WSSV) on different genes in order to investigate the immunological mechanisms behind the genetic resistance and to assess potential genes explaining disease resistance at the experimental level in the culture of Pacifc whiteleg shrimp (*Litopenaeus vannamei*) in Colombia. In Chile, the salmon microarray available for the consortium for genomics research on all salmon project (cGRASP[1]) in Canada has been used in collaboration with the University of Victoria for assessing disease resistance of piscirickettsia and infectious pancreatic necrosis virus (IPNV) in Atlantic salmon.

Genetic modifcation

A genetically modifed organism (GMO) is one whose genetic material has been altered through genetic engineering techniques with DNA molecules from different sources that are combined into one molecule to create a new set of genes. Typically, it involves introduction of a single gene from an unrelated species. After about two decades of very intensive research, the technology has reached the stage where it is possible to produce GM carp, tilapia and salmon. However, no aquatic GMOs have yet been approved for commercial release for food and agriculture purposes in any country. There are potential concerns about the environmental impact of raising such fsh (e.g. effects of possible interbreeding with native populations) and the greater amount of feed required for sustaining the increased growth rates, as well as problems with consumer acceptance, which may be one of the most important reasons that transgenic technology has not developed beyond the experimental phase. Many developing countries have yet to develop a clear policy on the use of transgenic fsh.

Molecular markers

Marker systems

Molecular markers are identifable DNA sequences found at specifc locations of the genome, transmitted by standard Mendelian laws of inheritance from one generation to the next. They rely on a DNA assay and a range of different kinds of molecular marker systems exist, such as restriction fragment length polymorphisms (RFLPs), random amplifed polymorphic DNAs (RAPDs), amplifed fragment length polymorphisms (AFLPs) and microsatellites. The technology has improved in the past decade and faster, cheaper systems like SNPs are increasingly being used. The different marker systems may vary in aspects such as their technical requirements, the amount of time, money and labour needed and the number of genetic markers that can be detected throughout the genome (reviewed in detail in FAO, 2007b). RAPDs and AFLPs have been used extensively in aquaculture due to their relatively easy development, i.e. they do not require construction of genomic libraries. Microsatellite markers are used increasingly in aquaculture species, due to their higher polymorphic information content, codominant mode of expression, Mendelian inheritance, abundance and broad distribution throughout the genome.

Molecular markers are being applied in developing countries in both aquaculture and fsheries management. Here, an overview is provided on their use for parentage analysis and genetic selection in aquaculture and for fsheries management and stock enhancement.

Parentage analysis

Molecular markers can be used successfully to trace alleles inherited by progeny from a group of candidate parents, thus providing a means of parentage analysis. In many fsh and shellfsh species, reproduction cannot be fully controlled and thus natural mating is the only way to produce offspring for the next generation of a breeding programme. For example, tilapia and carp breeding typically involves mass spawning where males and females are stocked in large "hapas" suspended in ponds, where a relatively large number of parents spawn simultaneously. Since constrained rates of inbreeding are required for sustained rates of genetic gain, in uncontrolled mating schemes it is not always possible to control the genetic contributions of broodstock or, therefore, the rates of inbreeding in a breeding programme using pedigree information. Small sample sizes together with sperm competition, mating preference (as in Artemia) and other biological factors after fertilization can increase the variance of family size, thereby decreasing the N_e to unsustainable levels.

When it is possible to control matings, one of the most important constraints still facing effective breeding programmes of species such as salmon, carp and trout is that newborn individuals are too small to be tagged individually using the traditional marking systems for livestock. The application of sustainable breeding programmes requires tagging a constant number of individuals from each family with passive integrated transponders (PIT tags) when they become suffciently large after a period of individual family rearing, in order to manage the rates of inbreeding. However, this system of early management creates common environmental effects for full-sib families.

To address these issues, mixtures of equal-aged progeny from different families can be reared communally to preclude the development of such family-specifc environmental effects, and genetic markers can be used subsequently to assign individuals to families after evaluation of individual performance. Thus, the impact of early common environmental effects is considerably reduced if markers are used for parentage analysis when selecting individuals for early growth rate traits. Several multinational salmon companies are using this system of tagging but there is still no information regarding its economic value compared with conventional tagging systems such as PIT tags. This may be important in species such as carp and tilapia where the costs of genotyping can greatly outperform the use of tanks and individual tagging systems. Furthermore, it is expected that rates of genetic gain for economic traits will not be signifcantly affected when common environmental effects are present.

Even though there is a plethora of information in the scientifc literature on the use of markers for parentage analysis in fsh and shellfsh, this procedure has not been fully used in species such as tilapia in developing countries where basic conventional breeding programmes have proved very successful. The sample size (i.e. the numbers of individuals and markers required for accurately reconstructing the pedigree of a population) is a practical issue since not all individuals in a population can be genotyped for all markers available. The issue of sample size may also arise in species where physical tagging is not possible or not economically sound (e.g. shrimp or marine species), or when disease challenges (e.g. with infectious pancreatic necrosis) are carried out very early stages in the life cycle.

For most breeding programmes, physical tagging will prove effcient both in economic and biological terms to achieve acceptable rates of genetic gain while minimizing rates of inbreeding. Genetic marker technology can still be costly in developing countries for routine assignment of parentage, although these costs can be reduced using multiplex PCR technology in which more than one marker can be genotyped simultaneously in a single gel lane or capillary. This is especially the case when only DNA markers are used without physical tagging, since individuals must be re-typed when records for multiple traits are included in the selection criteria. When it is possible to isolate families, multistage selection offers the possibility of frst selecting individuals on a within-family basis directly from tanks or hapas (for traits infuenced by common environmental effects) and then selecting at a second stage for traits measured at harvest. This alternative would maintain the rates of gain while decreasing the costs associated with tagging, or even increase rates of gain, when recording traits such as body weight from tanks (within families) that can be carried out relatively inexpensively.

Marker-assisted selection

Molecular markers can also be used in genetic improvement through MAS, where markers physically located beside (or even within) genes of interest (such as those affecting growth rates in salmon) are used to select favourable variants of the genes. MAS is made possible by the development of molecular marker maps, where many markers of known location are interspersed at relatively short intervals throughout the genome and the subsequent testing for statistical associations between marker variants and the traits of interest. In this way, genes (called QTLs) thought to control quantitative traits (traits of agronomic importance controlled by many genes and many non-genetic factors, such as growth rate in fsh) can be detected.

MAS can enhance rates of genetic gain compared with conventional breeding for traits that are diffcult or expensive to measure or when the heritability is relatively low. So far, many QTLs have been identifed in different experiments involving trout, salmon, carp and tilapia, but the main problem with the actual use in MAS is to have enough replications or powerful experiments to validate that the QTLs detected in a given experiment are actually real, and are segregating across populations or crosses. Furthermore, many of the QTLs detected were discovered using dominant markers such as RAPDs which are very diffcult to replicate in different laboratories, basically due to the use of insuffcient sample sizes and failure to account for the presence of false positives. This outcome is explained by the fact that there is a lack of complete genome sequences for many of the species currently used in aquaculture in developing countries such as tilapia, carp and shrimp. This is an important practical issue, because without information from physical maps it may be diffcult to characterize the actual genes explaining the genetic variation explained by the QTLs. This situation refects the relatively high level of fnancial resources needed both to carry out a genome sequence project for many species used in aquaculture and to actually implement a MAS programme. This is a very important issue in developing countries where smallholders are less likely to have the fnancial revenue to allow breeding programmes that incorporate the use of molecular information. Although MAS is potentially useful for many cultured species, conventional breeding programmes may be more proftable in the short to medium term in developing countries in low-input environments.

The development of molecular markers and linkage maps can greatly help scientists to understand the different factors that infuence the expression of quantitative traits. A number of genetic linkage maps have been published in aquaculture, some of the most comprehensive being for rainbow trout, channel catfsh, tilapias, Japanese founder and mussels. In shrimp, recent mapping has demonstrated the nature of sex control in shrimp as WZ/ZZ like chickens and unknown until now. Still, in important species such as Indian major carps and Chinese carps, they have not been developed. There are a number of ways in which this information can be used, the difference between them being the level of resolution with which these factors can be mapped. For example, QTLs with major effects on quantitative traits are mapped using markers to track the inheritance of chromosomal regions in families or in inbred line crosses using the extent of linkage disequilibrium generated in the population.

In practice, the identification of genes influencing specific traits is achieved using a combination of genetic mapping (linkage and fne mapping) to localize the QTL to a small region on the chromosome under analysis, and candidate gene or positional cloning approaches are used to identify the genes within the QTL region. According to the literature survey, it appears that very little information has come from developing countries on such research issues.

In some cases, it is possible to use suffcient biochemical or physiological information to investigate the association between the quantitative genetic variation and the level of marker polymorphisms within specifc genes. Nevertheless, this approach requires a great amount of detailed information in order to choose which gene explains the greatest effect and to have suffcient power to detect the association. This information is starting to appear in the aquaculture literature from multinational projects such as cGRASP, but it is still scarce for other fsh species of interest in developing countries.

So far, QTL mapping in aquaculture using commercial populations has been carried out mainly in developed countries, mostly with single-marker analysis (microsatellites and AFLPs) and using relatively sparse linkage maps when interval mapping is used. In tilapia, the F_2 design and a four-way cross between different species of *Oreochromis* have been used for detecting QTLs affecting cold tolerance and body weight. In outbred populations of salmonids, QTLs that infuence body weight have been mapped.

Studies seeking linkage of markers to traits amenable to MAS, such as disease resistance, have begun to appear in the literature over the past few years. For example, QTLs for resistance have been mapped for IPNV in salmonids, infectious salmonid anaemia, infectious haematopoietic necrosis virus and stress and immune response and cold tolerance in tilapia. Also, Somorjai, Danzmann and Ferguson reported evidence of QTLs for upper thermal tolerance in salmonids, with differing effects in different species and genetic backgrounds. To date, there are no examples of the application of these QTLs in practical fsh and shellfsh breeding programmes in developed or developing countries.

Biosecurity and Disease Control

Like other farming systems, the aquaculture industry has been overwhelmed by a fair share of transboundary aquatic animal diseases caused by viruses, bacteria, fungi, parasites and other undiagnosed and emerging pathogens.

Disease has thus become a primary constraint to the culture of many aquatic species, impeding both economic and social development in many countries. As a result, there will be increasing demand for improved aquatic animal biosecurity, particularly addressing the emerging health problems based on risk analysis. Epidemiological studies generate the data required for risk analysis; biosecurity measures require good information for accurate assessment and this leads to appropriate risk management. Thus, biosecurity, risk analysis and epidemiology are highly interrelated. All are aimed at making good use of scientifc research for disease prevention, control and management.

Of equal importance is the need for fundamental information that characterizes diseases in aquaculture. Import risk assessment will of necessity set the risk as "high" when there are little data on modes of transmission, host susceptibility, tolerance to abiotic factors (e.g. temperature, salinity) and immune response elicited, for a particular pathogen under consideration. The clear, unambiguous and rapid detection and identifcation of potential pathogens using morphological and molecular diagnostic tools are of paramount importance prior to making decisions on the disease status of any aquaculture zone.

Although conventional disease control strategies focus largely on diagnosis and therapy, the prevention of disease through vaccination, immunostimulation, the use of probiotics and bioremediation in culture environments, nutritional improvements etc., has also been practised. Signifcant advances in these areas have been achieved using biotechnological approaches.

Given the taxonomic diversity of aquaculture species, there is also a need to develop better information on the response of these species to disease in order to develop management strategies for them. Biotechnology approaches are sometimes the only means by which tools for this can be developed.

Pathogen Screening and Disease Diagnostics

The control of disease outbreaks relies heavily on having rapid and accurate diagnostic tools available in order to detect and identify the pathogen causing mortality. DNA and RNA methods have been used extensively for detecting a number of viral and bacterial pathogens in aquaculture worldwide. The techniques rely upon the fact that each pathogen species carries a unique

DNA or RNA sequence that can be used for identifcation. The techniques offer high sensitivity and specifcity, and the commercial development of PCR primers and diagnostic kits allows rapid screening for a number of serious viral and bacterial infections and has direct application. Molecular-based techniques such as PCR also have applications in situations where the animal shows no antibody response after infection. For example, as molluscs do not produce antibodies, antibody-based diagnostic tests have limited application to pathogen detection in these species.

Considering the diffculties that developing countries may face in using advanced molecular diagnostics, and the importance of gradually improving national diagnostic capacities in developing countries, FAO recommended a three-level diagnostic process. This involves: feld observations and necropsy (Level I); laboratory observations, bacteriology and histopathology (Level II); and electron microscopy, molecular biology and immunology (Level III). In countries where Level II and Level III diagnostic capabilities are not found, initial disease screening is carried out using Level I gross clinical examination. Accompanied by histopathology, this has been the traditional method of detecting pathogens in both developed and developing countries. There is a clear need to improve national diagnostic capacities to reach Level II and Level III diagnostic procedures, including molecular diagnostics.

These tools include both immunoassay- and DNA-based diagnostic methods, e.g. fuorescent antibody tests, enzyme-linked immunosorbent assays (ELISA), radioimmunoassay (RIA), *in situ* hybridization (ISH), dot blot hybridization and PCR amplifcation techniques. They are currently used to screen and/or confrm the diagnosis of many signifcant pathogens of cultured fnfsh such as channel catfsh virus, infectious haematopoietic necrosis virus, IPNV, viral haemorrhagic septicaemia virus, viral nervous necrosis virus and bacterial kidney disease, as well as shrimp diseases such as WSSV, yellow head virus (YHV), infectious hypodermal and haematopoietic necrosis virus (IHHNV) and Taura syndrome virus (TSV). Similar tools are under development for molluscan pathogens (*Haplosporidium* sp., *Bonamia ostreae*, *Marteilia refringens* and Herpes virus). Immunoassays and nucleic acid assays provide quick results with high sensitivity and specifcity at relatively low cost, and are particularly valuable for infections that are diffcult to detect (e.g. subclinical infections) using standard histology and tissue culture procedures. Molecular tools are also useful for research into the pathology and immunology of specifc infections. They can be used with non-lethal sampling and are valuable for monitoring challenge experiments under controlled laboratory conditions. Further development of these technologies is likely to speed up the detection (feld monitoring and laboratory examination) and diagnosis of disease, which is crucial for early and effective control of emergent disease situations.

Antibody-based techniques

A variety of antibody-based tests and molecular tests have been developed to detect mainly bacterial and viral fsh pathogens, although tests have also recently been reported for parasites and fungal agents. The antibody-based tests include slide agglutination, co-agglutination/latex agglutination, immunodiffusion, direct and indirect fuorescent antibody tests, immunohistochemistry and ELISA, dot blot/dip-stick and Western blot. The antibody-based test selected for the identifcation of pathogens depends on a variety of factors since each method has its merits and disadvantages. Although such methods are useful for the detection of pathogens in pure culture or/and in infected fsh tissue, their sensitivity thresholds limit their use in environmental samples, especially where pathogen levels are extremely low. DNA detection methods, however, such as PCR and ISH are ideally suited.

DNA-based techniques

Molecular technologies are also widely used for the detection of fsh pathogens. They have been successfully utilized for the detection and identifcation of low levels of aquatic pathogens. Such methods are also particularly useful for micro-organisms that are diffcult to culture, may exist in a dormant state, are involved in zoonosis, or in the elucidation of pathogen life cycles. In addition, molecular methods can be used for the identifcation of pathogens to the species level and in epidemiology for the identifcation of individual strains and differentiating closely related strains. Because of the general unavailability of the traditional pathogen isolation methods and immunodiagnostics for molluscs and crustaceans, molecular techniques have increasingly been used.

DNA-based methods such as PCR are extremely sensitive. However, false positive and false negative results can cause problems due to contamination or inhibition. Real-time PCR (closed tube to reduce contamination) and nucleic acid sequence-based amplifcation are alternatives that reduce these risks and offer high sample throughput. Some of the most common PCR-based technologies used for the detection of pathogens are nested PCR, RAPDs, reverse transcriptase PCR (RT-PCR), reverse cross blot PCR and RT-PCR enzyme hybridization assay. ISH is also widely used in the detection of shrimp viruses and in the confrmation of mollusc parasites. Colony hybridization has also been used successfully for the rapid identifcation of *Vibrio anguillarum* in fsh and has the advantage of detecting both pathogenic and environmental strains.

In recent years, the use of PCR-related tools has gained wide acceptance in developing countries. The advent of PCR has lead to important advances in the development of routine diagnostic tests, and it has been possible to develop probes aimed at the detection of pathogen genetic material in host tissue, as well as for assessing genetic variability within and between fsh and shellfsh populations. Both DNA- and RNA-based methods have been devised to detect pathogen genetic material. Depending on the pathogen, conventional PCR can be replaced by the more sensitive nested PCR method, in which primers within the region amplifed in a frst step are used for further amplifcation of DNA. RNA quantifcation can be carried out using RT-PCR of the viral nucleic acids present in sample tissues. As with the immunological methods described earlier, it should be noted that PCR does not demonstrate the presence of disease nor of a viable pathogen, but only that pathogen genetic material was present in the sample being examined. Despite this limitation and other problems related to ease of contamination, false positives, the limited number of primers available etc., when properly applied, PCR offers a relatively rapid and inexpensive way for the routine screening of large numbers of aquatic animals for commercial aquaculture and for testing of imported stocks during quarantine. For example, PCR is very important in the routine screening of massive numbers of penaeid shrimp larvae for serious viral pathogens such as WSSV, TSV etc. in Asian and Latin American countries.

DNA probes and epidemiology

DNA probes have particular value in the felds of epidemiology, routine disease surveillance and monitoring, treatment and eradication programmes in aquaculture and efforts to prevent the spread of pathogens to new geographical areas. These biotechnologies also have important application in risk management for aquatic animal diseases including inspection and certifcation of production and facilities and consignments for freedom from specifc pathogens; achieving recognition of a country as having disease-free status; and implementing disease zoning programmes and effective quarantine measures etc..

The Manual of Diagnostic Tests for Aquatic Animals, regularly published by the World Organisation for Animal Health (OIE), validates the use of traditional diagnostic methods such as evaluation of clinical signs, necropsy, histopathology, parasitology, bacteriology, virology, mycology etc., as well as immunological tests such as ELISA for the presumptive and confrmatory identifcation of OIE-listed diseases. The introduction to the Manual notes that "For the most part, molecular methods for fsh diseases are recommended for either direct detection of the pathogen in clinically diseased fsh or for the confrmatory identifcation of a disease agent isolated using the traditional method. With one or two exceptions, molecular techniques are currently not acceptable as screening methods to demonstrate the absence of a specifc disease agent in a fsh population for the purpose of health certifcation in connection with international trade of live fsh and/or their products. There is a need for more validation of molecular methods for this purpose before they can be recommended in the Aquatic Manual". This highlights the importance of further validating these diagnostic tools for serious and emerging diseases across a range of different laboratories worldwide.

Vaccines

Adams *et al.* reviewed the vaccine technologies in aquaculture. Vaccination is the action in which a host organism is exposed to organic (biological) molecules that allow the host to mount a specifc immune reaction through which it has a better capability to fght subsequent infections of a specifc pathogen compared with genetically similar non-vaccinated hosts. It has also been shown to be cost-effective and has led to the reduction in use of antibiotics. In Norway, for example, antibiotic use has decreased from 47 tons to approximately one ton annually.

A wide range of commercial vaccines is available against bacterial and viral pathogens and many new vaccines are under development. Most target salmon and trout, and there are expanding opportunities for marine fsh. Traditionally, the organic molecules used for vaccination are directly derived from the pathogen in question. The most straightforward approach is to culture the pathogen after it has been inactivated and presented to the host. So far, vaccines containing more than ten bacterial pathogens and fve viral pathogens have been produced based on such inactivated antigens. Alternatively, the pathogen is not inactivated but chemically or genetically weakened so as to survive only for a limited period in the host where it induces a specifc immune response without causing disease and mortality. Such vaccines are generally described as "live" vaccines, and there is concern that the attenuated strain may back-mutate and revert to the virulent wild type. Due to environmental and control concerns in most countries, only two live bacterial (*Edwardsiella ictaluri* and *Flexibacter columnarae* for Channel catfsh in the United States) and one live viral vaccine (koi herpesvirus for carp in Israel) are commercially available at present.

One of the most important factors leading to reduced antibiotic use by the aquaculture sector is the availability of good prophylactic measures for diseases causing severe mortalities in cultured fsh and shellfsh. The use of vaccines provides good immunoprophylaxis for some of the most important infectious diseases of fnfsh. In developed countries, their use has proved very effective at decreasing the unsustainable use of antibiotics. For example, in Norway antibiotic use in salmon farming has become almost negligible, at less than 1 gram per tonne of production, due mainly to the availability of vaccines for furunculosis and cold water vibriosis. At almost similar production levels, Chilean salmon farming shows much more antibiotic use due to the emergence of *Piscirickettsia salmonis*, a pathogen causing severe losses of stock prior to harvest. Thus, there have been recent attempts to develop immunoprophylactic measures.

As molecular-based vaccine production procedures rely heavily on biotechnological tools, vaccines are produced mainly in developed countries. A DNA vaccine is a circular DNA plasmid that contains a gene for a protective antigenic protein from a pathogen of interest. Considerable industrial research has been conducted towards developing DNA vaccines for species such as salmonids against pathogens (generally viruses) for which traditional methods have not been successful. As many strains and varieties of a single pathogen are generally present in the tropics, unlike in temperate pathogens, monovalent vaccines are not practical under tropical conditions. Such diffculties, together with the lack of adequate biotechnological knowledge and fnancial resources, have led to fewer advances in vaccine development in the tropics, and for tropical species. Commercial vaccines using inactivated bacterial pathogens are available for some species: channel catfsh, European seabass and seabream, Japanese amberjack and yellowtail, tilapia, Atlantic cod, salmon and trout. Fewer commercially available viral vaccines have been produced, and no commercially available parasite vaccines exist.

Environmental Management and Bioremediation

Aquaculture, like any other live production system, produces effuents rich in nutrients. Some aquatic production systems also produce effuents with harmful substances such as residues and metabolites of antibacterials and therapeutics. Developing systems that produce effuents with acceptable standards and improving the quality of the aquatic environment where effuent discharges are unacceptably high is a challenge. Biotechnological interventions such as bioremediation, the use of probiotics, and vaccination offer signifcant promise for addressing these important issues.

Bioremediation is a promising biotechnological approach for the degradation of hazardous waste to environmentally safe levels using aquatic micro-organisms or other fltering macro-organisms. Although this procedure has been used in various situations such as sewage treatment, application to shrimp and other aquaculture wastes is fairly novel. There are many commercial products on the market, mainly bacterial preparations, but the mode of action and effcacy of many of these have yet to be scientifcally measured. In addition to microbes, bivalves, seaweeds, holothurians (sea cucumbers) etc., have been tested to assess their ability to reduce organic loading or reduce the excess nutrients produced during culture production. Various bioremediation preparations have also been developed with a view to removing nitrogenous and other organic waste in water and bottom sludge and thus reduce chemically-induced physiological stress, e.g. in pond-reared shrimp. More products will undoubtedly emerge with continued research in this feld, but controlled feld trials are urgently needed to determine the effectiveness and cost-beneft of these products under culture conditions.

Probiotics are generally administered as live microbial feed supplements which affect the host animal by improving the intestinal microbial balance to optimize the presence of non-toxic species. A stable gut microfora helps the host resist pathogenic invasions, particularly via the gastro-intestinal tract. Antibiotics reduce specifc or broad-spectrum gut microfora and probiotics may have post-antibiotic treatment potential for restoring the microbial balance. Probiotics are widely used in animal husbandry but their use in aquaculture is still relatively new. However, there are increasing reports of potential probiotics for shrimp aquaculture which has been plagued by opportunistic bacteria such as the luminescent *Vibrio harveyi*, and in some cases probiotics have been reported to signifcantly reduce antibiotic use in shrimp hatcheries. Suppression of proliferation of certain pathogenic bacteria (e.g. *Vibrio* spp.) in shrimp hatcheries has been achieved by introducing (inoculating) non-pathogenic strains or species of bacteria that compete for microbial metabolite resources. This procedure shows promise to be effective and economical. However, further refnement of the administration and concentration loads needed for effective pathogen suppression is required. Effective and economically viable probiotics also require greater research into optimal strains of probiotic micro-organisms and stringent evaluation under feld conditions.

The control of disease using vaccines is a reputed technology. There are interesting examples of reducing antibacterial use in aquaculture through the use of vaccination particularly in temperate species such as salmon and trout. Reduction of the use of antibacterials not only diminishes the risk of rejection of aquatic products at international trading borders due to the presence/detection of residues above acceptable levels, it also helps in reducing the contamination of natural water bodies with harmful residues and the development of antimicrobial-resistant bacteria.

The proliferation of red tides with the blooming of harmful algae has been increasingly reported in many parts of Latin America, where the toxins represent a threat to food safety as well as a cause of fsh and shellfsh losses from the associated mortalities. Red tides can produce signifcant economic losses to fsheries and aquaculture due to bans on the marketing of fsh and shellfsh from the affected geographical area and to the toxic effects on fsh. In Central America and the Caribbean, la “ciguatera” is the most important cause of toxic poisoning, resulting from consumption of tropical fsh. In Latin America, blooms of *Alexandrium* spp. are one of the major causes of large economic losses due to the banning of commercial sales of mussels. In Chile, preventive closures cause about US$100 million in annual losses to the artisanal bivalve fshery. Furthermore, these closures have a direct negative impact on local employment in the shellfsh production sector, which is labour intensive, thus having a detrimental effect on livelihoods. While it is not known if climate change is increasing the number of episodes of algal blooms, it is recognized that red tide episodes have recently become more common. Warm episodic currents also play a key role in causing large economic losses through mass mortalities of fsh.

To date, the detection of toxins due to algal blooms is carried out using mouse bioassays and high performance liquid chromatography, but new methodologies are being developed for detection of *Alexandrium catenella.* Expressed sequence tag (EST) libraries are now publicly available, so that it may be possible to develop molecular diagnostic techniques. To improve the prevention of impacts on aquaculture, PCR techniques and EST libraries can be used also to assist the early detection of toxin-producing algae in vast marine areas.

Biodiversity Conservation and Fsheries Management

Restocking procedures are common in many developing countries, but the potential of restocking and stock enhancement stems primarily from the development of the technologies used to produce hatchery-reared juveniles. The production of large numbers of juveniles and their subsequent release into the wild can affect a fshery resource in at least two ways : 1) when stocking is done to restore a spawning biomass there is some scope for interbreeding between the natural population and the introgressed population and 2) there may be enough individuals used to restore the carrying capacity of the fshery.

From a genetic point of view, the main consequence of restocking may be the hybridization of non-native individuals with natural stocks, which can have important impacts on natural biodiversity. Fish are very prolifc, and under many hatchery production systems a relatively small number of parents can provide suffcient numbers of juveniles for release, in which case the genetic variability of the fshery may be reduced. This situation can easily lead to genetic bottlenecks, the forthcoming generations of population being subjected to relatively high rates of inbreeding thus inadvertently reducing the genetic variability of the population. This can have large effects on the sensitivity of individuals to environmental variations and could possibly cause the extinction of a population or species in a particular environment. In addition, inbreeding can affect growth and reproduction.

The mating of wild fsh with those released by restocking programmes can promote the loss of genes important for local adaptation in a genetic mechanism called outbreeding depression. While this concern has been effectively studied in terrestrial animals and in salmon populations in developed countries, this is not the case in other fsheries from developing countries.

Therefore, careful restocking procedures need to be developed in order to reduce the potential for the introgressed population to reduce the genetic variability and therefore the sustainability of the resource. Assessing the genetic diversity of managed stocks or highly selected populations is an important issue when pedigree information is lacking or in situations where some kind of quality assurance is needed.

The use of molecular markers and the principles of population genetics have proved very effective in assessing the actual levels of genetic variability within single populations and in measuring the extent of differentiation between populations. For example, the Centro Nacional de Pesquisa de Peixes Tropicais in Brazil has studied the use of RAPD markers for the Amazonian fsh "matrinxa" (*Brycon cephalus*) and has shown a relatively large reduction in genetic variability in fsh used for restocking purposes compared with the native Amazonian river population.

In developing countries, the markers have been used mainly for assessing genetic variation in tilapia and carp populations in Thailand, the Philippines and India. Markers have been used for characterizing stocks and comparing levels of genetic variability in *Oreochromis* species. Agustin used markers to assess genetic differences between indigenous samples from Africa and populations from Asia, concluding that the low performance of *O. mossambicus* stocks can be explained by the effect of large bottlenecks in the populations used for aquaculture in Asia. Molecular markers have also been used to assess population differentiation of Nile tilapia (*O. niloticus*) for both domesticated and feral populations. In both cases, moderate to great genetic differentiation was found between strains and the use of markers successfully correlated with the actual biogeographical data.

The escape of farmed fsh from aquaculture may infuence the genetic variability of native populations. The possible genetic impacts resulting from introductions and invasive alien species include: interbreeding between alien and native genotypes causing, in some cases, reduced reproductive effciency and generating nonviable offspring; decreased ftness from loss of co-adapted gene complexes; and indirect genetic impacts resulting from other ecological interactions.

Climate change and related climatic events such as the El Niño-Southern Oscillation (ENSO) can have serious impacts on the distribution of fshery resources between countries. Based on census numbers, mackerel fsheries were apparently depleted in Chilean coastal waters during the occurrence of ENSO episodes. However, markers have shown little differentiation with other populations in the Pacifc Ocean (such those observed in New Zealand), and so it is likely that the drop in numbers is related to migration of the mackerel populations to colder waters in the Pacifc rather than to fshery depletion.

Compared with livestock and crop production, aquaculture is a novel production system in many developing and developed countries. As shown above, biotechnologies are being applied in fsheries management but their use is very limited compared with aquaculture. The use of successful and effective biotechnologies in aquaculture is very much confned to genetic manipulations and improvements, and to health management.

The success or failure in using biotechnologies in developing countries depends to a large extent on: 1) the markets for each of the products within the production sectors, and 2) the investment and acquisition capacity for the fsheries and aquaculture sectors. In the case of aquaculture, the latter is very important considering that the largest proportion of world production comes from developing countries and from small farmers (specifcally in Asia). Most biotechnological interventions have been developed for improved production and the better management of aquaculture. Most have been targeted towards high value commercial aquaculture species generally produced for international markets. Although many small-scale farmers are producing for export markets, the signifcant uptake of many biotechnological interventions and innovations has generally been restricted to commercial or industrial aquaculture operations. This is certainly due to the cost of the technologies as well as the organized nature of industrial aquaculture.

Recently, however, as a result of better organization in the small-scale farming sector, certain biotechnologies have been effectively taken up by the small farmers in many parts of the developing world. They include DNA probes for detecting pathogens in some species (mainly PCR detection of major viral pathogens of shrimp), the use of SPF shrimp broodstock or postlarvae, the use of certain DNA vaccines, the all-male (genetically male) tilapia and, in some cases, markers for pedigree evaluation in salmon worldwide. In fact, almost everywhere in the world, shrimp farmers, whether small or large, currently use only PCR-tested postlarvae for stocking. For example, in India there are more than 90 laboratories providing PCR services for the shrimp sector – mainly for the screening of seed and broodstock. In Vietnam, there are over 40 laboratories. This pattern holds true in many countries of the region as the cost of using such biotechnologies has declined over the years and the benefts have increased tremendously.

The majority of aquaculture produce comes from the small-scale farming sector, in many instances comprising low-input extensive production systems. Although there is scope for biotechnologies, and although they are already being employed by small-scale farmers, classical environmental improvements and better management practices such as conventional genetic selection of broodstock, conventional health management through the avoidance of pathogens etc., can also contribute signifcantly towards improving small-scale aquaculture production and sustainability.

Case Studies

Biotechnologies are used in aquaculture for reducing losses due to diseases and improving production through genetic manipulation. These technologies are regularly used in almost all countries at different rates and levels based on the intensity and commerciality of the production system. Here, two case studies are presented, outlining specifc successful applications of biotechnological tools in aquaculture in developing countries.

PCR-based Pathogen Detection in Shrimp Aquaculture in India

At present, shrimp is the most valuable aquaculture commodity sector in the world. This sector has been continuously facing the challenge of new diseases, particularly viral pathogens. Some 20 years ago, there was hardly any accurate molecular-based pathogen detection system available in any part of the world. Now, as a result of advanced molecular research and biotechnology, there are many DNA-based detection technologies such as PCR methodologies available for all the major shrimp viruses. A number of PCR, nested-PCR and hybridization tests have been developed for virus detection. The tests use a range of different PCR primers and hybridization probes targeted to different and poorly defned sites in the virus genome. Several RT-PCR tests are also available. The application of PCR detection of viruses of broodstock and postlarvae in both *Penaeus monodon* and *Penaeus vannameii* is now practised in all countries producing commercial shrimp at all levels. Recently, lateral fow chromatographic immunodiagnostic strips similar to common drug store pregnancy tests have begun to appear for some shrimp diseases. Using these, unskilled farm personnel can easily diagnose shrimp disease outbreaks at the farm. The strips are relatively cheap and quick. Other methods comparable to PCR and RT-PCR are now available or are being developed for single and dual or multiple viral detection but they currently require advanced equipment and personnel.

This rapid detection technology has given a new dimension to the shrimp industry and losses due to viral diseases have been reduced tremendously by the use of PCR-tested postlarvae for stocking. Recent successes in farmer group or cluster formation and management in shrimp aquaculture, particularly in India and Indonesia, are to a large extent based on good health management which includes the use of PCR tested postlarvae for stocking in ponds. This demonstrates a scenario in which a successful biotechnology has not only contributed towards realizing its scientifc objective, but also towards improving the overall governance of the sector.

To consider a specifc case study, the use of PCR detection technology was the key basic step towards developing an effective better management practice (BMP) for small-scale shrimp aquaculture in Andhra Pradesh. In India, aquaculture is mainly carried out by small- and marginal-scale farmers located in the remote villages of the country. They are largely unorganized, scattered and poorly educated. The farmers mostly opt for traditional methods for operating their farms and do not have access to technological innovations or scientifc applications. A joint MPEDA-NACA (Marine Products Export Development Authority – Network of Aquaculture Centres in Asia-Pacifc) project assisted by FAO was initiated in 2002 to support shrimp farmers in disease control and coastal management, leading to the participatory development of BMPs that provided signifcant improvements in profts and reduced shrimp disease risks for farmers. One of the key interventions that the farmers adopted in applying BMPs in their quest to reduce losses due to disease was the use of PCR-screened postlarvae for stocking.

The project supported farmers in the implementation of BMPs through the formation of self-help groups around local "clusters". An economic analysis of 15 farmer groups in Andhra Pradesh clearly demonstrated that farmers adopting BMPs including the use of PCR-screened postlarvae for stocking had higher proftability, lower production costs and were able to produce quality and traceable shrimp without using any banned chemicals.

The project has been highly successful in forming a self-help movement of farmers across India through a grassroots approach. From a mere fve farmers who frst adopted the cluster-farm approach and BMPs in 2002, the programme had swelled to more than 1 000 farmers in 30 aquaculture societies in fve coastal states by 2007. Beginning in 2007, the MPEDA-NACA project became the National Centre for Sustainable Aquaculture (NaCSA). NaCSA is an outreach organization of MPEDA established to service the small-scale aquaculture sector and provide technical support to farmer groups. It aims to empower and build the capacity of small-scale farmers to produce quality shrimps in a sustainable and more proftable manner.

Perhaps one of the keys to the above success is the ability to reduce losses due to disease in production systems, and to a large extent this has been possible through the use of PCR technology for screening and detecting major viral pathogens in broodstock and postlarvae.

Specifc Pathogen-free Stocks in Shrimp Aquaculture

Only a few species have so far been domesticated in the aquaculture sector. One group of species on which most research has been focused on the domestication and development of SPF strains is the penaeid shrimp. SPF shrimp are produced in SPF facilities using many biotechnological tools, particularly DNA-based pathogen detection and diagnostic techniques. The primary goal of SPF facilities is to produce strains of shrimp that are disease-free, domesticated and genetically improved for aquaculture. SPF lines are available for *P. vannamei*, *P. stylirostris* and *P. monodon*. The SPF status should signify that the shrimp have passed through a rigorous quarantine and disease-screening process that has found them to be free from specifed pathogens of concern to culturists. This characteristic means that countries or regions which still do not have this species can be reasonably sure that importation of SPF animals will not result in the introduction of the specifed pathogens from which the animal is declared free. This does not, however, guarantee against the animal being infected with unknown pathogens or known pathogens for which the animal was not screened.

Genuine SPF shrimp are produced in biosecure facilities that have been repeatedly examined and found free of specifed pathogens using intensive surveillance protocols, and originate from broodstock developed with strict founder population development protocols. These founder populations are generated by extensive quarantine procedures that result in SPF F_1 generations derived from wild parents. Only stocks raised and held under these conditions can be considered truly SPF. There is not yet an internationally agreed protocol for the development of SPF shrimp, and certainly some variation exists in the quality of different SPF stocks. Once the animals are removed from the SPF production facilities, they should no longer be referred to as SPF even though they may remain pathogen-free. Once outside the SPF facility, the shrimp may be designated as High Health (since they are now subject to a greater risk of infection), but only if they are placed into a well-established facility with a history of disease surveillance and biosecurity protocols. If the shrimp are put elsewhere, for example into a non-biosecure maturation unit, hatchery or farm, they can no longer be called SPF or High Health as they are now exposed to a high risk of infection.

One potential drawback of SPF animals is that they are only SPF for the specifc diseases for which they have been checked. Typically this will consist of the viral pathogens which are known to cause major losses to the shrimp culture industry, including WSSV, YHV, TSV, IHHNV, *Baculovirus penaei* virus and Hepatopancreatic parvovirus as well as microsporidians, haplosporidians, gregarines, nematodes and cestodes. Despite this screening, new, hidden or "cryptic" viruses may be present, but because they are as yet unrecognized they may escape detection. Thus, it is believed that SPF shrimp shipped from Hawaii resulted in the contamination of shrimp in Brazil and Colombia with TSV. This was because, at the time, TSV was not known to have a viral cause and therefore went unchecked in SPF protocols.

In any case, the use of SPF stocks is only one part of a complete plan for minimizing disease risks in shrimp culture. The development of SPF strains is really designed to ensure that postlarvae stocked into grow-out ponds are free of disease, which is one, if not the most serious, source of contamination. Other areas of this strategy that must be implemented include ensuring that broodstock, eggs, nauplius, larvae and juveniles derived from SPF stock remain SPF.

Creating an enabling public sector environment is essential to improve governance at all levels of aquaculture development. There have been many regulatory rebounds in the aquaculture sector, in particular in shrimp farming in some countries. Uncontrolled and unregulated development of the sector has outstripped the carrying capacity in some locations, causing signifcant production losses mainly due to disease and resulting in the complete abandonment of farms. Signifcant improvements have been made in mitigating such catastrophic problems, and the negative environmental and social impacts of shrimp farming throughout the world have been signifcantly reduced. The use of wild-caught postlarvae in shrimp culture, which has a signifcant impact on aquatic biodiversity, has almost stopped or is little practised. The recent development of SPF broodstocks of some species of shrimp has reduced reliance on wild-caught postlarvae to a minimum.

SPF shrimp if produced and maintained under good biosecurity have proved successful. The success of SPF stocks may be more pronounced in large-scale industrial shrimp culture facilities where maintaining stringent biosecurity is possible. The use of this successful biotechnological approach in the rather disorganized small-scale shrimp aquaculture production sector poses another challenge.

Preparing for the Future

Key Issues where Biotechnologies Could be Useful

Environmental sustainability

Aquaculture is the fastest growing food producing sector in the world. It is poised to expand, diversify and intensify over the coming decades to bridge the increasing global gap between the supply and demand of aquatic food. Responsible production through sustainable practices is the key to achieving this massive task. In the effort to maximize the contribution from aquaculture it is inevitable that many constraints and hurdles need to be overcome. The biggest hurdle is to maintain environmental sustainability,

Conventional methods of controlling diseases such as chemotherapeutants are ineffective for many new pathogens (notably viruses). Molecular techniques have therefore received increasing attention for pathogen screening and identifcation. In addition, these biotechnologies are providing signifcant insights into pathogenesis (disease development) and show strong potential for disease control and prevention programmes (e.g. DNA vaccines), as well as for treatments of diseases. The increased sensitivity and specifcity conferred by DNA- or RNA-based probes has provided signifcant inroads for the early detection of diseases and identifcation of subclinical carriers of infections. This has had a direct effect on enhancing preventative management and control of disease in cultured species. Concomitant with this has been a decrease in the need for reactive treatments using traditional methodologies such as antibiotics or culling and disinfection. This has been particularly successful for shrimp broodstock selection and has broken the infection cycle perpetuated for years by accidental broodstock transmission of viral pathogens to developing offspring.

Biotechnologies can provide much assistance to improve aquatic animal health management in aquaculture in developing countries, in particular through the development of sensitive and accurate molecular diagnostic methods and tools as well as vaccines for tropical diseases. Bioremediation and probiotics also provide some further opportunities.

Climate change

In the future, one of the greatest constraints could be the impact of climate change on aquaculture. Climate change threatens fsheries and aquaculture through higher temperatures and changes in weather patterns, water quality and supply. Important differences in the magnitude and types of impacts on aquaculture are predicted for different regions.

The ability to adapt will confer a major advantage and should be developed by countries and regions. There is a need for the aquaculture sector to join other economic sectors in preparing to address the potential impact of global warming. One of the practical responses to climate change for aquaculture could be to strengthen the adaptive capacity and resilience of the sector, particularly those of small farmers and aquatic resources users. Increased resilience is a desirable feature of any sector. It can mitigate the future impact of unforeseen events (e.g. economic change, disease epidemics, tsunamis, etc.), including those related to climate.

There is some knowledge and experience from aquaculture itself, and from the broader area of agriculture and natural resource management, which could be used. Aquaculture, and particularly mariculture, could in fact provide adaptation opportunities to produce good quality protein when freshwater may become scarce. On the other hand, freshwater aquaculture can produce protein with higher water saving than other animal production sectors. Certain biotechnologies, particularly those dealing with genetic improvement, health and environmental mitigation should be of signifcant value for the discovery of adaptive technologies and interventions to counter the ever-present menace of climate change.

Identifying Options for Developing countries

To bridge the future gap between demand and supply of aquatic food, production needs to be almost double in less than three decades. In the quest to meet this unprecedented demand, the aquaculture sector will face serious constraints. Four major constraints are inevitable: 1) disease prevention and health management, 2) genetic improvement and domestication, 3) environmental management and 4) food safety. These constraints are not new. They have been constantly addressed during the development of aquaculture over the past two decades, including through the use of biotechnologies.

Over the years, aquaculture biotechnologies and other technological innovations have had a positive impact on aquaculture diversification, investment potential, and international technology exchange. The development of biotechnologies in aquaculture should therefore provide a means of producing healthy and fast-growing animals by environmentally friendly means. However, this development will largely depend on the desire and willingness of producers to work hand-in-hand with scientists, and on the international donor community's readiness to assist developing counties in the related research, capacity building and infrastructure development. Improved exchange of information and discussion between scientists, researchers and producers from different regions about their problems and achievements will undoubtedly help this important sector to develop with a view to increasing sustainable global aquatic animal production.

Based on the overview and analysis contained in this Chapter, a number of specifc options can be identifed for developing countries to help them make informed decisions regarding the adoption of biotechnologies in the future, such as when – and if – they should deploy one or more biotechnologies and, if they decide to do so, how they can ensure the successful application of the chosen biotechnologies to enhance food security in the future.

— Few biotechnological advancements and tools are currently in use in small-scale aquaculture operations aiming at rural development, poverty alleviation and food security. However, there is a need to identify these, their application and socio-economic impact in developing countries. Developing countries should therefore collect information on the aquatic animal biotechnologies that may be used and analyse their national-level adoption and socio-economic impacts. Such information should be used to advise policy-makers on the cost/beneft implications of such applications. Increased efforts should be made to develop aquatic biosecurity policies within national research and development (R&D) programmes or national aquatic production programmes.

— The use of biotechnologies in aquaculture worldwide has increased incrementally over the past two decades. Several aquaculture biotechnologies have been used for improving aquatic food production in both developed and developing countries and have significant potential for future improvement. Since most aquaculture biotechnologies are still too technical and costly for small-scale farmers, efforts should be made to develop low-cost simple technologies that are easy to introduce to less advanced aquaculture farmers. Developing countries should give priority to developing aquaculture biotechnologies which are appropriate and conducive for both industrial and small-scale farmers.

— Major biotechnological achievements and advances in fsheries and aquaculture have been mainly restricted to aquaculture and to the fields of genetics, health and the environment. Genetic improvements using genetic manipulation and hormonal therapy etc. have shown promise for producing fsh and shellfsh with improved and desirable production qualities. Disease prevention and health management in aquaculture have benefted signifcantly from advances in biotechnologies. Many reliable and accurate rapid diagnostic techniques have been developed which can be used by small-scale farmers. There are several effcient vaccines now available for certain aquaculture species which have signifcantly reduced the use of antibacterials in their culture. However, more research is required to develop vaccines for tropical species, particularly the major species of global production. Some environmental remediation tools and technologies have been developed using several biotechnologies. They are being applied in some production systems but their broad adoption across different production systems and practices is yet to be established.

The potential contribution of biotechnologies for genetic improvement to improve production of culture aquatic species should be recognized. National research and development plans should include appropriate research in these areas. In aquatic animal health research, the development of molecular diagnostics, vaccines and probiotics should be prioritized and national research institutions should also carry out research using appropriate biotechnologies that can help the development of sustainable aquaculture in this area. National governments embarking on aquaculture development should also recognize that there is ample evidence for positive aquatic environmental impacts using various biotechnological interventions, and therefore the use of biotechnology for improving the aquatic environment should be considered.

— Until recently, perhaps because the application of biotechnologies in fsheries and aquaculture has been mainly restricted to the commercial and industrial aquaculture of temperate species, there has been little evidence in many developing countries of national-level efforts to prioritize the development and application of biotechnologies in aquaculture. Even when efforts were made to develop such technologies in the public sector of developing countries, they were not always directed towards or made available to improve small-farmer livelihoods. There is a need to create national policy environments in developing countries, including suitable investment and funding opportunities, to allow the development and application of appropriate biotechnologies in support of aquaculture development. National governments should pay special attention to the small-scale aquaculture sector. Preferential treatment of the sector towards capacity building in appropriate biotechnologies should also be considered.

— The funding required in developing countries for aquatic biotechnological research and applications should be found through national budgets or through extra budgetary resources. An integral part of funding should be directed towards investment in capacity building in the relevant felds of the aquaculture sector. A suitable investment environment and funding opportunities should be created to allow the development and application of appropriate biotechnologies in support of aquaculture development. The appropriate involvement of the relevant stakeholders in decision-making processes should be assured.

— The establishment of effcient institutional structures and enforceable legal frameworks are important for the responsible use of biotechnologies in aquaculture at the national level. Such institutional arrangements should also strengthen research and extension needs and enhance relevant human and infrastructural capacities. National legal frameworks in aquaculture biotechnologies should be developed within an integrated national biotechnology framework, which also complies with the legal or voluntary requirements of international treaties and agreements that the country has ratifed.

— National biotechnology programmes in developing countries should include a special committee to oversee the aquatic biotechnology programme and research. Such committees should be formed in all countries and regional cooperation should be sought.

— Information gathering and dissemination on aquatic biotechnologies should be encouraged within and between countries in a given region, and developing countries should consider setting up dedicated websites for this purpose.

— Aquaculture products are facing increasing competition in accessing international markets. One of the key criteria is food safety and compliance with international food safety standards. Many such standards can be met through better farming that uses both simple and advanced biotechnological interventions. The aquaculture industry should therefore consider the importance of such biotechnological interventions in improving and maintaining food safety of cultured aquatic products. National governments in developing countries should consider R&D interventions on food safety within the broader framework of biotechnology.

Identifying Priorities for Action for the International Community

The international community, including FAO and other UN organizations, NGOs, donors and development agencies, can play a key role in supporting developing countries by providing a framework for international cooperation and funding support for the generation, adaptation and adoption of appropriate biotechnologies.

Here, a set of Priorities for Action is identifed that can assist the international community in playing this role.

— Relevant international institutions, donors and development partners should recognize that biotechnological interventions can contribute to sustainable aquaculture development worldwide.

— Relevant international agencies should assist developing countries to collect, collate and analyse information about the biotechnologies in use in fsheries and aquaculture, and their contributions to national food security, poverty alleviation and social development.

— Relevant international agencies should make efforts to maintain databases and information systems to assist countries access information for national biotechnology development programmes relating to fsheries and aquaculture.

— Donors and international funding agencies supporting sustainable aquaculture development for food security and poverty alleviation should dedicate an appropriate share of their assistance projects to promoting and strengthening aquatic biotechnology R&D in developing countries. International research efforts should focus on developing interventions that are accessible to small-scale farmers.

— When supporting the application of biotechnologies in fsheries and aquaculture, the international community should consider that technical assistance in biotechnology R&D should not be done at the expense of funding for other key research felds and that it should support effective and intimate links to strong breeding and extension programmes.

— The international community assisting developing countries towards aquaculture sustainability should consider biotechnological advancement as an important area to be supported, and should assist developing countries in strengthening capacities for biotechnology policy development and long-term planning.

— The international community should assist developing countries to develop the capacities of their national agricultural research systems, which include aquaculture, to involve relevant stakeholders in decision-making processes.

— The international community should assist developing countries in establishing adequate institutional capacities for the development and enforcement of regulations related to use of biotechnologies in fsheries and aquaculture.

References

Adams, A. & Thompson, K.D. 2006. Biotechnology offers revolution to fish health management. *Trends Biotechnol.,* 24: 201205.

FAO. 2003. *Recent technological innovations in aquaculture*, by R.P. Subasinghe, D. Curry, S.E. McGladdery & D. Bartley. *In* Review of the state of world aquaculture, pp. 5974. FAO Fisheries Circular No. 886. Rome.

Liu, Z.J. & Cordes, J.F. 2004. DNA marker technologies and their applications in aquaculture genetics. *Aquaculture,* 238: 137.

Subasinghe, R.P., Soto, D. & Jia, J. 2009. Global aquaculture and its role in sustainable development. *Rev. Aquac.,* 1: 29.

Zohar, Y. & Mylonas, C.C. 2001. Endocrine manipulations of spawning in cultured fish: From hormones to genes. *Aquaculture,* 197: 99136.

Bibliography

Adams, A. & Thompson, K.D. 2006. Biotechnology offers revolution to fish health management. *Trends Biotechnol.,* 24: 201205.

Ambrosia, R.E. and De Wall, D.T. 1990. Diagnosis of parasitic disease. *Reviews of Sci Techn., Office Intern. Epizool.* 9, 759-778.

Beardmore, J.A.B. 1997. Transgenics: autotransgenics and allotransgenics. *Transgenic Research* 6: 107-108.

Bondari, K., and R.A. Dunham. 1986. The effect of inbreeding on reproduction, growth and survival of channel catfish. *Aquaculture* 57:363–364.

Bruce D. and A. Bruce, 1999. *Engineering Genesis: The Ethics of Genetic Engineering,* London: Earthscan Publications.

Carpenter, J.E. & Gianessi, L.P. 2001. *Agricultural biotechnology: updated benefits estimates.* Washington, DC, National Center for Food and Agricultural Policy.

Doyle, R.W., and A. Talbot. 1986. Effective population size and selection in variable aquaculture stocks. *Aquaculture* 57:27–35.

Falconer, D.S., 1960. *Introduction to quantitative genetics.* London, Oliver and Boyd

FAO. 2003. *Recent technological innovations in aquaculture*, by R.P. Subasinghe, D. Curry, S.E. McGladdery & D. Bartley. *In* Review of the state of world aquaculture, pp. 5974. FAO Fisheries Circular No. 886. Rome.

Fjalestad, K.T., Gjedrem, T. & Gjerde, B. 1993. Genetic improvement of disease resistance in fish: An overview. *Genetics in Aquaculture* IV. Eds G. A. E. Gall and H. Chen pp.65-74.

Friedman, Yali. 2008. *Building Biotechnology: Starting, Managing, and Understanding Biotechnology Companies.* Washington, DC: Logos Press.

Gatesoupe, F. J., 1999. The use of probiotics in aquaculture: a review. *Aquaculture* 180: 147-165.

Gjoen, H.M. & Bentsen, H.B. 1997. Past, present and future of genetic improvement in salmon aquaculture. *ICES Journ. Mar. Sci.* 54: 1009-1014.

Gopakumar, K. 2003. Indian aquaculture. *Journal of Applied Aquaculture,* 13(1/2): 1-10.

Hardy, R.W. 2006. Removing persistent organic pollutants from fish meal and oil. *Aquaculture Magazine* 32(1): 43-45.

Helsop-Harrison, J., 1974. Genetic resources conservation: the end and the means. *J.R.Soc.Arts,* Feb. issue:157 p.

Hoban, T. 2004. *Public attitudes towards agricultural biotechnology*. ESA Working Paper, forthcoming. Rome, FAO.

Houdebine, L.M. 2003. *Animal transgenesis and cloning*. New York, John Wiley and Sons.

Kincaid, H.L. 1983. Inbreeding in fish populations used for aquaculture. *Aquaculture* 33:215–227.

Kirpichnikov, V.S. 1981. *Genetic Bases of Fish Selection.*Springer-Verlag, New York, New York, USA.

Knowles, D.P. and Gorham, J. R. 1990. Diagnosis of viral and bacterial diseases. *Reviews of Sci. Techn., Off. Intern. Epizool.* 9, 733-757.

Lightner, D.V. and Redman, R.M. 1998. Shrimp diseases and current diagnostic methods. *Aquaculture* 164, 201-220.

Liu, Z.J. & Cordes, J.F. 2004. DNA marker technologies and their applications in aquaculture genetics. *Aquaculture,* 238: 137.

Maitland, P.S., 1979. The status and conservation of rare freshwater fishes in the British Isles. *Proc.Brit. Freshwat.Fish Conf.*, 1:237–48

Moav, R., T. Brody and G. Hulata, 1978. Genetic improvement of wild fish populations. Science, *Wash.*, 201(2):1090–4

Muller M., Kersten, S., 2003. Nutrigenomics: Goals and Perspectives. *NatureReviews Genetics* 4. 315 -322.

National Research Council (NRC). 2002. *Animal biotechnology: science-based concerns*. Washington, DC, National Academies Press.

Oliver, Richard W. *The Coming Biotech Age.*

Pew Initiative on Food and Biotechnology. 2003. *Future fish: issues in science and regulation of transgenic fish.* Washington, DC, Pew Initiative on Food and Biotechnology. 72 pp.

Pingali, P & Raney, T. 2003. *Globalization and agricultural biotechnology: impacts and implications for developing countries.* ESA Working Paper. Rome, FAO.

Plumb, J.A. 1997. Trends in freshwater fish disease research. p. 35-47. *In:* T.W. Flegel and I.H. MacRae (eds.) *In:Diseases in Asian Aquaculture* III. Fish Health Section, Asian Fish. Soc., Manila.

Pray, C.E. & Naseem, A. 2003. *Biotechnology R&D: policy options to ensure access and benefits for the poor.* ESA Working Paper 03-08. Rome, FAO.

Rana, K.J. 1997. Recent trends in global aquaculture production: 1984-1995. 14-19p. FAO *Aquaculture Newsletter.* No. 16. August 1997, FAO, Rome. 28pp.

Sagar, A., Daemmrich, A. & Ashiya, M. 2000. The tragedy of the commoners: biotechnology and its publics. *Nature Biotechnology*, 18 (January 2000): 2-4.

Subasinghe, R.P., Soto, D. & Jia, J. 2009. Global aquaculture and its role in sustainable development. *Rev. Aquac.,* 1: 29.

Tave, D. 1993. *Genetics for Fish Hatchery Managers,* 2nd ed. Van Nostrand Reinhold, New York, New York, USA.

Thieman, W.J.; Palladino, M.A. 2008. *Introduction to Biotechnology*. Pearson/Benjamin Cummings.

Thompson, P.B. 1997. *Food biotechnology in ethical perspective.* London, Blackie Academic & Professional.

Traxler, G. 2004. *Economic impacts of biotechnology-based technological innovations*. ESA Working Paper, forthcoming. Rome, FAO.

Villaluz, D.K., 1966. *The Lake Lanao fisheries and their conservation.* Manila, Bureau of Printing, 53 p.

Zhu, Z. 1992. Generation of fast growing transgenic fish: Methods and mechanisms. In *Transgenic Fish.* Eds. C. L. Hew and G. L. Fletcher. Singapore, Singapore World Scientific, pp.92-119.

Zohar, Y. & Mylonas, C.C. 2001. Endocrine manipulations of spawning in cultured fish: From hormones to genes. *Aquaculture,* 197: 99136.